T0353480

Set Theory and Foundations of Mathematics

An Introduction to
Mathematical Logic

Volume II
Foundations of Mathematics

Set Theory and Foundations of Mathematics

An Introduction to
Mathematical Logic

Volume II
Foundations of Mathematics

Douglas Cenzer
University of Florida, USA

Jean Larson
University of Florida, USA

Christopher Porter
Drake University, USA

Jindřich Zapletal
University of Florida, USA

 **World Scientific**

NEW JERSEY · LONDON · SINGAPORE · BEIJING · SHANGHAI · HONG KONG · TAIPEI · CHENNAI · TOKYO

Published by

World Scientific Publishing Co. Pte. Ltd.

5 Toh Tuck Link, Singapore 596224

USA office: 27 Warren Street, Suite 401-402, Hackensack, NJ 07601

UK office: 57 Shelton Street, Covent Garden, London WC2H 9HE

Library of Congress Cataloging-in-Publication Data

Names: Cenzer, Douglas, author. | Larson, Jean (Jean Ann), author. |
 Porter, Christopher (Christopher P.), author. | Zapletal, Jindřich, 1969–   author.
Title: Set theory and foundations of mathematics : an introduction to mathematical logic /
 Douglas Cenzer, Jean Larson, Christopher Porter, Jindřich Zapletal.
Description: New Jersey : World Scientific, [2020] | Includes bibliographical references and index. |
 Contents: Volume. I. Set theory
 --
Identifiers: LCCN 2020007602 | ISBN 9789811201929 (v. 1 ; hardcover) |
 ISBN 9789811201936 (v. 1 ; ebook)
Subjects: LCSH: Set theory--Textbooks. | Logic, Symbolic and mathematical--Textbooks.
Classification: LCC QA248 .C358 2020 | DDC 511.3/22--dc23
LC record available at https://lccn.loc.gov/2020007602

Volume II: Foundations of Mathematics
ISBN 978-981-124-384-4 (hardcover)
ISBN 978-981-124-385-1 (ebook for institutions)
ISBN 978-981-124-386-8 (ebook for individuals)

British Library Cataloguing-in-Publication Data

A catalogue record for this book is available from the British Library.

Copyright © 2022 by World Scientific Publishing Co. Pte. Ltd.

All rights reserved. This book, or parts thereof, may not be reproduced in any form or by any means, electronic or mechanical, including photocopying, recording or any information storage and retrieval system now known or to be invented, without written permission from the publisher.

For photocopying of material in this volume, please pay a copying fee through the Copyright Clearance Center, Inc., 222 Rosewood Drive, Danvers, MA 01923, USA. In this case permission to photocopy is not required from the publisher.

For any available supplementary material, please visit
https://www.worldscientific.com/worldscibooks/10.1142/12456#t=suppl

Desk Editors: Jayanthi Muthuswamy/Lai Fun Kwong

Typeset by Stallion Press
Email: enquiries@stallionpress.com

Printed in Singapore

Preface

This book was developed over many years from class notes for a course on the foundations of mathematics at the University of Florida. This course has been taught to advanced undergraduates as well as lower level graduate students. The notes have been used more than thirty times as the course has been developed. The companion volume on set theory has a similar history.

Set theory and mathematical logic make up the foundation of pure mathematics. The companion volume [4] on set theory focuses on the axioms of set theory and their applications. Using the axioms of set theory, we can construct our universe of discourse, beginning with the natural numbers, moving on with sets and functions over the natural numbers, integers, rationals and real numbers, and eventually developing the transfinite ordinal and cardinal numbers. Mathematical logic provides the language of higher mathematics which allows one to frame the definitions, lemmas, theorems and conjectures which form the every day work of many mathematicians. The axioms and rules of deduction set up the system in which we can prove our conjectures, thus turning them into theorems. A major theme is the connection between mathematical logic and other areas of mathematics, including algebra, analysis, geometry and topology.

Gödel's Incompleteness Theorem is also a significant focus of this volume.

It is reasonable to cover most of the material in a one semester course, with selective omissions. Here is an outline, with some topics indicated to be optional. The core of this volume consists of Chapters 2, 3, 5 and 6. Chapter 1 is a review of propositional logic, including the language and the deductive calculus, along with notions of soundness, completeness, and compactness. This chapter should be covered as needed in two or three weeks, with Section 1.8 on logical vs. topological compactness being optional. Chapter 2 presents the language and the deductive calculus of predicate logic, and introduces the study of structures or models, in about two weeks. Chapter 3 covers the Completeness Theorem, and the notions of compactness, isomorphism and elementary equivalence, in two or three weeks. The part of Section 3.4 on persistence is optional, as is Section 3.5 on categoricity and quantifier elimination. Computability is developed in Chapter 5, including Turing machines, recursive functions and the Halting Problem, in two or three weeks, with Section 5.1 on finite state machines optional. The notion of decidability is covered in Chapter 6, leading to Gödel's Incompleteness Theorem, in about two weeks. If all goes well, this leaves three or four weeks for the other chapters. Chapter 4 is a detailed development of Boolean algebras as far as ultraproducts. Chapter 7 introduces the relatively new area of algorithmic randomness, via Kolmogorov complexity. Chapter 8 covers nonstandard natural and real numbers. One approach uses ultraproducts, but this can be omitted if Section 4.4 has not been covered. Chapter 9 covers the foundations of geometry.

The applications of logic found in Chapters 4, 7, 8, and 9 give the instructor options for creating a course which makes a solid connection with other areas of mathematics.

The book contains nearly 100 examples and 200 exercises, which test the students understanding and also enhance the material.

We would like to thank our colleagues, students and postdocs who have read versions of this book and provided many helpful comments. This includes Ethan McCarthy and Tom Winckler in particular. We have enjoyed teaching from these notes and are very pleased to share them with a broader audience.

About the Authors

Douglas Cenzer is Professor of Mathematics at the University of Florida, where he was Department Chair from 2013 to 2018. He has to his credit more than 100 research publications, specializing in computability, complexity and randomness. He joined the University of Florida in 1972 after receiving his Ph.D. from the University of Michigan.

Jean Larson is Emeritus Professor at the University of Florida, specializing in combinatorial set theory. She received her Ph.D. in mathematics from Dartmouth University in 1972 and was E.R. Hedrick Assistant Professor at UCLA from 1972 to 1974, before joining the University of Florida in 1974.

Christopher Porter is Associate Professor and Chair of Computer Science, and also holds a position in Mathematics, at Drake University, specializing in computability theory, algorithmic randomness, and the philosophy of mathematics. He received his Ph.D. in mathematics and philosophy from the University of Notre Dame in 2012, was an NSF international postdoctoral fellow at Université Paris 7 from 2012 to 2014, and a postdoctoral associate at the University of Florida from 2014 to 2016.

Jindřich Zapletal is Professor of Mathematics at University of Florida, specializing in mathematical logic and set theory. He received his Ph.D. in 1995 from the Pennsylvania State University, and held postdoctoral positions at MSRI Berkeley, Cal Tech and Dartmouth College, before joining the University of Florida in 2000.

Contents

Chapter 1

Introduction

The goal of this book and its companion volume [4] on set theory is to provide an introduction to mathematical logic and the foundations of mathematics for students of mathematics. Mathematical logic is a flourishing area of research with specialties including computability theory, model theory, proof theory, and set theory, and all of these areas are covered. An effort is made to connect foundations with topics covered in the mathematics major, including algebra, analysis, geometry, and topology. Thus we have chapters on Boolean algebras, on nonstandard analysis, and on the foundations of geometry. We hope that this book can provide a broader perspective for all students interested in mathematics, including prospective logic students.

Chapter 1 begins with a review of propositional logic, including the language of propositional logic and its deductive calculus. The language of propositional logic has propositional variables and a number of connectives, including symbols for "and", "or", "not", and "implies". The notion of rank is defined for propositions and used to prove some properties using induction. Truth interpretations (truth tables) provide models for the sentences of propositional logic. The Disjunctive Normal Form Theorem shows that all propositions can be represented using only negation, disjunction, and conjunction. The Soundness Theorem and the Completeness Theorem are proved for the deductive system, as well as the Compactness Theorem. The notions of completeness, consistency, and independence of a theory are presented. The natural topology on the set of truth interpretations is introduced and logical compactness is compared with topological compactness.

Chapter 2 introduces predicate logic. The symbols of predicate, or first-order, logic include quantifiers, relation symbols, constant symbols, and function symbols, as well as the propositional connectives. The language has terms, interpreted by elements of a structure, as well as formulas. The notion of satisfaction of a formula by a model is presented. A natural deduction system is given in which the rules of inference are closely connected to the connectives and quantifiers. Soundness of the proof system is shown, that any provable sentence is satisfied by every model.

Chapter 3 is an introduction to model theory, beginning with the Completeness Theorem, that any sentence which is satisfied by every model can be proven. The Compactness Theorem is proved, as well as the Löwenheim–Skolem Theorem. The notions of isomorphism and elementary equivalence of structures are presented. The concepts of axioms and axiomatizability are introduced and it is shown that the notion is finiteness is not axiomatizable. The concepts of universal and existential formulas are given and some notions of persistence are studied. Categoricity and quantifier elimination are also studied, including the Łoś-Vaught Test. The theory of infinity is shown to have quantifier elimination and the theory of dense linear orderings without endpoints is shown to be countably categorical.

Chapter 4 is a detailed development of Boolean algebras as far as ultraproducts. Axioms are given for Boolean algebras with operations of meet, join, and complement. The partial ordering on a Boolean algebra is also studied, along with the notions of atomic and atomless Boolean algebras. The notions of filters and ideals are given, including principal, prime, and maximal ideals. This leads to ultrafilters and ultraproducts. Examples of Boolean algebras include power sets, interval algebras, and Lindenbaum algebras of sentences.

Computability theory is developed in Chapter 5, including Turing machines, recursive functions, and the Halting Problem. Finite state machines (FSMs) and also finite state transducers are introduced as the first model of computation. A function is Turing computable if there is a Turing machine which computes it. Here a language is the set of strings accepted by a machine. Regular languages are those accepted by FSMs, and semicomputable languages are those accepted by Turing machines. Examples are given of functions which can be computed by finite state transducers and functions which cannot be so computed. Examples are given of languages which are regular and languages which are not regular. Primitive recursive and partial recursive functions are defined. It is shown that the Turing computable functions are the same as the partial recursive functions, an illustration of the Church–Turing Thesis. The Halting Problem is defined and shown to be semicomputable but not computable.

The notions of decidability and undecidability are covered in Chapter 6, leading to Gödel's Incompleteness Theorem. It is shown that any effectively axiomatizable theory is decidable. An effective quantifier elimination procedure for a theory also shows that the theory is decidable. Examples include the theory of infinity and the theory of dense linear orderings without endpoints. Axioms for Peano Arithmetic (PA) are given and it is shown that the computable functions are exactly those definable in PA. This is used to prove Church's Theorem that PA is not decidable. This further leads to a proof of Gödel's result that PA is not complete.

Chapter 7 introduces the relatively new area of algorithmic randomness, via Kolmogorov complexity. The Kolmogorov complexity $C(\sigma)$ of a string σ is defined and shown to be invariant up to a constant. The notion of incompressibility is defined and it is shown that incompressible strings exist. The complexity function C is shown to be not computable. We also prove Chaitin's Theorem, that for some $L \in \mathbb{N}$, PA cannot prove that $C(\sigma) \geq L$ for any string σ, which provides an alternative proof of the incompleteness of Peano Arithmetic.

Chapter 8 covers nonstandard natural and real numbers. Compactness implies the existence of models $\mathbb{N}^*$ of arithmetic elementarily equivalent to $\mathbb{N}$ which contain *infinite* natural numbers and of models $\mathbb{R}^*$ equivalent to $\mathbb{R}$ which contain *infinitesimals*. Such models may also be constructed using ultraproducts. Concrete models of certain subsets of the Peano axioms are presented. The use of infinitesimals in the calculus is examined in the notions of limits, continuity, and derivatives.

Chapter 9 covers the foundations of geometry. A geometry is a system with points and lines satisfying certain axioms. Various parallel postulates, specifying the number n of lines which are parallel to a given line and passing through a particular point. Classically, it was conjectured that the standard parallel postulate ($n = 1$) was a consequence of the other axioms of Euclidean geometry, but this was much later shown to be false. The models of Lobachevsky (where $n = \infty$) and Riemann (where $n = 0$) are presented. There is also a section on finite geometries, projective planes in particular. In teaching our course on foundations, we sometimes cover this chapter first, as it presents concrete examples of the notions of axioms and independence, and since Euclidean geometry was among the first areas to be studied from a foundational point of view.

Chapter 2

Propositional Logic

2.1. Basic Definitions

Propositional logic concerns relationships between sentences built up from primitive proposition symbols with logical connectives. The history of propositional logic traces back to the classical era and was put in symbolic form during the 19th century by logicians George Boole [3] and Augustus De Morgan [9], in particular.

Definition 2.1.1. For any set Σ of symbols, a *string* from Σ is a finite sequence $a_1, a_2, \ldots, a_n$ of symbols from Σ. The set of all strings from Σ is denoted by Σ^*. In the case that $n = 0$, we have the empty string ε. The set of symbols is sometimes referred to as an *alphabet* and strings are also known as *words*.

For example, the standard English alphabet has 26 symbols and words from the English language include some of the words from this alphabet.

The symbols of the language of propositional calculus are as follows:

(1) logical connectives: $\neg$, $\wedge$, $\vee$, $\rightarrow$;
(2) punctuation symbols: (,);
(3) propositional variables: $A_0, A_1, A_2, \ldots$.

A propositional variable is intended to represent a proposition which can either be true or false. We can specify restricted propositional languages, $\mathcal{L}$, by specifying a subset of the propositional variables. In this case, let $\mathrm{PVar}(\mathcal{L})$ denote the propositional variables of $\mathcal{L}$.

Definition 2.1.2. The collection of *sentences*, denoted by $\mathrm{Sent}(\mathcal{L})$, of a propositional language $\mathcal{L}$ is the smallest set S of strings in $\mathcal{L}$ such that

(1) each propositional variable in the set $\mathrm{PVar}(\mathcal{L})$ is in S;
(2) the set S is closed under the following production rules:

    (a) If $A \in S$, then so is $(\neg A)$.
    (b) If A and B are in S, then so is $(A \wedge B)$.
    (c) If A and B are in S, then so is $(A \vee B)$.
    (d) If A and B are in S, then so is $(A \to B)$.

Notice that as long as $\mathcal{L}$ has at least one propositional variable, then $\mathrm{Sent}(\mathcal{L})$ is infinite. When there is no ambiguity, we will drop parentheses. For example, $(A_1 \wedge A_2)$ may be shortened to $A_1 \wedge A_2$ and $(\neg(\neg A_3))$ may be shortened to $\neg\neg A_3$.

Sentences may be ranked by examining their construction under Definition 2.1.2.

Definition 2.1.3.

(1) Each propositional variable A in the set $\mathrm{PVar}(\mathcal{L})$ has rank 0, that is $\mathrm{rk}(A) = 0$.
(2) The notion of rank can be extended to arbitrary sentences as follows. Let A and B be sentences with $\mathrm{rk}(A) = m$ and $\mathrm{rk}(B) = n$.

    (a) $\mathrm{rk}((\neg A)) = m + 1$.
    (b) For any binary connective $*$ from $\{\wedge, \vee, \to\}$, $\mathrm{rk}((A * B)) = 1 + \max\{\mathrm{rk}(A), \mathrm{rk}(B)\}$.

We let Sent_n be the set of sentences of rank n.

Each sentence has a derivation under Definition 2.1.2 and a corresponding rank. The derivation is obtained by breaking down the sentence into its component parts.

Example 2.1.4. The sentence $(((A_0 \vee (\neg A_1)) \wedge A_2) \rightarrow A_3)$, which we will write as B, may be derived as follows. First we see that B can be written as

$$(B_0 \rightarrow A_3),$$

where B_0 is $((A_0 \vee (\neg A_1)) \wedge A_2)$. Hence $\mathrm{rk}(B) = 1 + \max\{\mathrm{rk}(B_0), \mathrm{rk}(A_3)\}$. Next we see that B_0 can be further written as

$$(B_1 \wedge A_2),$$

where B_1 is $(A_0 \vee (\neg A_1))$, so that $\mathrm{rk}(B_0) = 1 + \max\{\mathrm{rk}(B_1), \mathrm{rk}(A_2)\}$. Then B_1 can also be written as $(A_0 \vee B_2)$, where $B_2 = (\neg A_1)$, so that $\mathrm{rk}(B_1) = 1 + \max\{\mathrm{rk}(A_0), \mathrm{rk}(B_2)\}$. Finally $\mathrm{rk}(B_2) = 1 + \mathrm{rk}(A_1)$.

Now we reverse this process to calculate the rank of B. First we note that $\mathrm{rk}(A_0) = \mathrm{rk}(A_1) = \mathrm{rk}(A_2) = 0$, since these are propositional variables. Thus $\mathrm{rk}(B_2) = 1 + 0 = 1$ and $\mathrm{rk}(B_1) = 1 + \max\{0, 1\} = 1 + 1 = 2$. So $\mathrm{rk}(B_0) = 1 + \max\{2, 0\} = 3$, and finally, $\mathrm{rk}(B) = 1 + \max\{3, 0\} = 4$.

We can define functions on sentences by recursion on rank, and prove theorems about $\mathrm{Sent}(\mathcal{L})$ by induction on rank.

Definition 2.1.5. For any symbol a in $\mathcal{L}$ and any sentence in $\mathrm{Sent}(\mathcal{L})$, let $\#_a(P)$ be the number of occurrences of a in P. For a subset S of symbols, let $\#_S(P)$ be the number of occurrences of elements of S in P, that is, $\#_S(P) = \sum_{a \in S} \#_a(P)$.

The following lemma may be used to prove some properties of sentences.

Proposition 2.1.6. *The number $\#_a(P)$ of occurrences of a in a propositional sentence P can be defined recursively on $\mathrm{Sent}(\mathcal{L})$, as follows:*

(1) *Let P be the propositional variable A_i.*
  (a) *If a is A_i, then $\#_a(P) = 1$.*
  (b) *If a is any other symbol, then $\#_a(P) = 0$.*
(2) *Let P be $(\neg Q)$.*
  (a) *If a is $\neg$, or is a left or right parenthesis, then $\#_a(P) = \#_a(Q) + 1$.*
  (b) *If a is any other symbol, then $\#_a(P) = \#_a(Q)$.*
(3) *Let P be $(Q * R)$, where $*$ is a binary connective.*
  (a) *If a is $*$, or is a left or right parenthesis, then $\#_a(P) = \#_a(Q) + \#_a(R) + 1$.*
  (b) *If a is any other symbol, then $\#_a(P) = \#_a(Q) + \#_a(R)$.*

Example 2.1.7. It is clear that the number of left and right parentheses in a propositional sentence must be equal. Here is how to prove this, using course-of-values induction on the rank n of a sentence.

Base Step: $n = 0$. If P has rank 0, then P is a propositional variable, so that $\#_((P) = 0 = \#_)(P)$.

Induction Step: $n > 0$. Let Q have rank $n > 0$ and suppose by induction that $\#_((P) = \#_)(P)$ for all sentences of rank $< n$. There are two cases.

Case 1: Q is $(\neg P)$ for some P of rank $< n$. Then by induction $\#_((P) = \#_)(P)$. Thus

$$\#_((Q) = \#_((P) + 1 = \#_)(P) + 1 = \#_)(Q).$$

Case 2: Q is $(P * R)$ for some P, R of rank $< n$, and some binary connective $*$. Then by the induction hypothesis, $\#_((P) = \#_)(P)$ and $\#_((R) = \#_)(R)$. Thus

$$\#_((Q) = \#_((P) + \#_((R) + 1 = \#_)(P) + \#_)(R) + 1 = \#_)(Q).$$

Here is a more challenging result. Let us say that an initial segment Q of a string P is *proper* if it is not empty and also is not equal to P.

Proposition 2.1.8. *For any propositional sentence P and any proper initial segment Q of P, $\#_((Q) > \#_) (Q)$.*

Proof. The proof is by induction on the rank of P.

Base Step: If P is a propositional variable, then it has no proper initial segments.

Induction Step: Let P have rank n. We consider two cases.

Case 1: P is $(\neg P_1)$ for some P_1 of rank $n-1$. There are three possibilities for a proper initial segment Q. The first two are that Q is (or that Q is $(\neg$, and then $\#_((Q) = 1 > 0 = \#_) (Q)$.

Otherwise, $Q = (\neg Q_1$, where either Q_1 is P_1 or Q_1 is an initial segment of P_1. It follows from the inductive hypothesis that $\#_((Q_1) \geq \#_) (Q_1)$ (where the equality is for the case that Q_1 is P_1), and so we have

$$\#_((Q) = \#_((Q_1) + 1 \geq \#_) (Q_1) + 1 = \#_) (Q) + 1 > \#_) (Q).$$

Case 2: $P = (P_1 * P_2)$ for some P_1, P_2 of rank $< n$ and some binary connective $*$. Then, by the induction hypothesis, any proper initial segment Q_1 of P_1 has more left parentheses than right parentheses, and similarly for a proper initial segment Q_2 of P_2. Here are the possibilities for a proper initial segment Q of P. First we may have that Q is (, so as above $\#_((Q) = 1 > 0 = \#_) (Q)$. Second, we may have that Q is $(Q_1$, where Q_1 is a proper initial segment of P_1, and then

$$\#_((Q) = \#_((Q_1) + 1 > \#_) (Q_1) = \#_) (Q).$$

Third, we may have that Q is $(P_1$ or Q is $(P_1 *$, and then

$$\#_((Q) = \#_((P_1) + 1 = \#_) (P_1) + 1 > \#_) (P_1) = \#_) (Q).$$

Fourth, we may have that Q is $(P_1 * Q_2$, where either Q_2 is P_2 or is a proper initial segment of P_2. As in the second part of Case 1, it follows from the inductive hypothesis that $\#_((Q_2) \geq \#_)(Q_2)$, so that we have

$$\#_((Q) = \#_((P_1) + \#_((Q_2) + 1 \geq \#_)(P_1) + \#_)(Q_2) + 1$$
$$= \#_)(Q) + 1 > \#_)(Q). \qquad \square$$

Corollary 2.1.9. *For any propositional sentences P and Q, if Q is a proper initial segment of a sentence P, then Q is not a sentence.*

It is very important that a sentence be unambiguous. The results above about parentheses play a key role here. Note that if we just write $A \vee B \wedge C$, then this might be read as $A \vee (B \wedge C)$, or as $(A \vee B) \wedge C$. The first sentence is a disjunction, whereas the second sentence is a conjunction, and these two sentences should not be confused with each other. In fact, the string $A \vee B \wedge C$ does not belong to Sent$(\mathcal{L})$. Randomly chosen strings of symbols are not expected to be sentences.

The following proposition will help us to see whether or not a given string is a sentence.

Theorem 2.1.10 (Unique Readability). *Any sentence P in* Sent$(\mathcal{L})$ *is exactly one of the following conditions*:

(1) *a propositional variable*;
(2) *a negation $(\neg Q)$ with unique $Q \in$ Sent$(\mathcal{L})$*;
(3) *a disjunction $(Q \vee R)$ with unique $Q, R \in$ Sent$(\mathcal{L})$*;
(4) *a conjunction $(Q \wedge R)$ with unique $Q, R \in$ Sent$(\mathcal{L})$*;
(5) *an implication $(Q \to R)$ with unique $Q, R \in$ Sent$(\mathcal{L})$*.

Proof. It is immediate from Definition 2.1.2 that a sentence must have one of these forms. The uniqueness can be shown by induction on rank.

Base Step: If P is a propositional variable, then it has length 1, so it cannot have any other form.

Induction Step:

Case 1: Q is $(\neg P)$ for some P. If Q can also be written as $(\neg P_1)$, then clearly P and P_1 are the same propositional variable, so that P is unique. Now suppose that at the same time Q is $(Q_1 * Q_2)$ for some sentences Q_1, Q_2 and some binary connective $*$. This would mean that Q_1 begins with a $\neg$, which is impossible by Proposition 2.1.8.

Case 2: Q is $(P_1 * P_2)$ for some sentences P_1, P_2 and some binary connective $*$. It follows from Case 1 that Q cannot also be a negation. Now suppose that Q can also be written as $(R_1 \diamond R_2)$ for some sentences R_1, R_2 and some binary connective $\diamond$. Suppose by way of contradiction that $R_1 \neq P_1$. Then without loss of generality R_1 is an initial segment of P_1, so that R_1 is not a sentence by Corollary 2.1.9. Thus R_1 and P_1 are the same propositional variable and hence $*$ and $\diamond$ are the same propositional connective. It is now immediate that R_2 and P_2 are equal as well. □

Example 2.1.11. Let P be $(A_1 A_2 \vee A_3)$, where A_1, A_2, A_3 are sentences. How can we be sure that this is not a sentence? Clearly P is not a propositional variable and it is not a negation. Suppose therefore that $P = (Q * R)$ for some sentences Q, R and some binary connective $*$. As in the argument above, we must have that Q is A_1, or else one of them would be a proper initial segment of the other. But this means that A_2 begins with the connective $*$, which is clearly impossible.

It is possible to minimize the amount of parentheses in a sentence by introducing certain conventions, as in algebra. For example, the exterior parentheses can be deleted. Negation $\neg$ is presumed to apply to the lowest rank sentence that follows it, so that for example $\neg A \to B$ must mean $((\neg A) \to B)$. This is similar to reading $a - b \times c$ as $a - (b \times c)$ in arithmetic. Conjunction $\wedge$ has the next highest priority, followed by disjunction $\vee$. Thus $A \vee B \wedge C$ means $(A \vee (B \wedge C))$. This is similar to reading $a + b \cdot c$ as $a + (b \cdot c)$. Finally, implication $\rightarrow$ has the lowest priority. Thus the sentence $(((\neg A_2) \wedge (\neg A_3)) \rightarrow (\neg (A_2 \vee A_3)))$ may be written more simply as $\neg A_2 \wedge \neg A_3 \rightarrow \neg (A_2 \vee A_3)$.

Since both $\vee$ and $\wedge$ are associative, we can just write $A \wedge B \wedge C$ and likewise $A \vee B \vee C$ without ambiguity. However, some parentheses are still necessary. For example, $A \rightarrow (B \rightarrow C)$ is different from $(A \rightarrow B) \rightarrow C$ (in a sense we will make precise in the next section).

Exercises for Section 2.1

Exercise 2.1.1. Determine which of the following strings is a sentence, where A_0, A_1, A_2 are propositional variables.

(a) $((A_0 \rightarrow (\neg A_1 \vee (\neg A_2)))$;
(b) $(A_1 \rightarrow (A_1 \neg \wedge A_2))$;
(c) $(((A_1 \vee \neg A_2) \wedge A_3))$;
(d) $(((\neg A_2) \vee (\neg A_3)) \leftrightarrow (\neg (A_2 \wedge A_3)))$.

Exercise 2.1.2. Show that there are no sentences of length 2, 3 or 6, but that any other length is possible.

Exercise 2.1.3. Prove by induction on rank that in any sentence P, the number of occurrences of propositional variables equals one plus the number of occurrences of binary connectives.

Exercise 2.1.4. Prove by induction on rank that, for any n, and any sentence P of rank $n > 0$, P has an initial segment Q such that $\#_(Q) = n + \#_)(Q)$.

Exercise 2.1.5. Find a derivation of the sentence $B = (\neg((A_0 \wedge (A_1 \vee (\neg A_2)))) \rightarrow A_3))$ and use this to calculate the rank of B.

Exercise 2.1.6. Show that $(P_1 \neg P_2)$ is not a sentence for any sentences P_1, P_2.

Exercise 2.1.7. Use the conventions described above to simplify the sentence: $(((A \rightarrow C) \wedge (B \rightarrow C)) \rightarrow ((A \vee B) \rightarrow C))$.

Exercise 2.1.8. Put the parentheses back into the simplified sentence $A \vee \neg B \to C \wedge D$ to make it a legal sentence.

2.2. Truth Interpretations

In order to use propositional logic, we would like to give meaning to the propositional variables. Rather than assigning specific propositions to the propositional variables and then determining their truth or falsity, we consider truth interpretations.

Definition 2.2.1. A *truth interpretation* for a propositional language $\mathcal{L}$ is a function

$$I : \mathrm{PVar}(\mathcal{L}) \to \{0, 1\}.$$

If $I(A_i) = 0$, then the propositional variable A_i is considered to represent a false proposition under this interpretation. On the other hand, if $I(A_i) = 1$, then the propositional variable A_i is considered to represent a true proposition under this interpretation.

There is a unique way to extend the truth interpretation to all sentences of $\mathcal{L}$ so that the interpretation of the logical connectives reflects how these connectives are normally understood by mathematicians.

Definition 2.2.2. Let $I : \mathrm{PVar}(\mathcal{L}) \to \{0, 1\}$ be a truth interpretation for a propositional language $\mathcal{L}$. By abuse of notation, we also denote by I the unique extension of I to the collection of all sentences of the language. This is defined by recursion, using the following closure rules:

(1) If $I(A)$ is defined, then $I(\neg A) = 1 - I(A)$.
(2) If $I(A)$ and $I(B)$ are defined, then $I(A \wedge B) = I(A) \cdot I(B)$.
(3) If $I(A)$ and $I(B)$ are defined, then $I(A \vee B) = \max\{I(A), I(B)\}$.

(4) If $I(A)$ and $I(B)$ are defined, then

$$I(A \to B) = \begin{cases} 0 & \text{if } I(A) = 1 \text{ and } I(B) = 0, \\ 1 & \text{otherwise.} \end{cases}$$

The truth interpretation of $A \vee B$ above defines the *inclusive* or, since $A \vee B$ is true as long as at least one of $\{A, B\}$ is true. The *exclusive* or, denoted by $A + B$ is true when exactly one of $\{A, B\}$ is true.

Intuitively, tautologies are statements which are always true, and contradictions are ones which are never true. These concepts can be defined precisely in terms of interpretations.

Definition 2.2.3. A sentence P is a *tautology* for a propositional language $\mathcal{L}$ if for every truth interpretation I, $I(P) = 1$. P is a *contradiction* if for every truth interpretation I, $I(P) = 0$. Two sentences P and Q are *logically equivalent*, in symbols $P \Longleftrightarrow Q$, if every truth interpretation I takes the same value on both of them, that is, $I(P) = I(Q)$. A sentence P is *satisfiable* if there is some truth interpretation I with $I(P) = 1$.

The notion of logical equivalence is an equivalence relation; that is, it is a reflexive, symmetric and transitive relation. The equivalence classes given by logical equivalence are infinite for nontrivial languages (i.e., those languages containing at least one propositional variable). However, if the language has only finitely many propositional variables, then there are only finitely many equivalence classes.

Notice that if $\mathcal{L}$ has n propositional variables, then there are exactly $d = 2^n$ truth interpretations, which we may list as $\mathcal{I} = \{ I_0, I_1, \ldots, I_{d-1} \}$. Since each I_i maps the truth values 0 or 1 to each of the n propositional variables, we can think of each truth interpretation as a function from the set $\{0, \ldots, n - 1\}$ to the set $\{0, 1\}$. The collection of such functions can be written as $\{0, 1\}^n$, which can also be interpreted as the collection of binary strings of length n.

Definition 2.2.4. A *truth function* is a map from the set of truth interpretations to $\{0, 1\}$. For a fixed n, let $\mathcal{F}_n$ denote the set of truth functions mapping truth interpretations on n propositional variables to $\{0, 1\}$.

So a truth function can be viewed as a possible column of a truth table. Since there are $d = 2^n$ truth interpretations over n variables, there are exactly $2^d = 2^{2^n}$ many possible truth functions in $\mathcal{F}_n$. Each sentence φ gives rise to a function $TF_\varphi : \mathcal{I} \to \{0, 1\}$ defined by $TF_\varphi(I_i) = I_i(\varphi)$. Informally, TF_φ lists the column under φ in a truth table. Note that for any two sentences φ and ψ, if $TF_\varphi = TF_\psi$ then φ and ψ are logically equivalent. Hence there are at most 2^{2^n} equivalence classes of the sentences over n variables. We will see in the next section that every truth function corresponds to some propositional sentence, so that there are exactly 2^{2^n} equivalence classes.

Truth tables may be used to demonstrate that given sentences are logically equivalent. Some examples are provided by the following lemma (equivalences (2) and (3) are known as *De Morgan's Laws*). This result will be needed to prove the Disjunctive Normal Form Theorem.

Lemma 2.2.5. *The following pairs of sentences are logically equivalent as indicated by the metalogical symbol $\Longleftrightarrow$:*

(1) $\neg\neg A \Longleftrightarrow A$.
(2) $\neg A \vee \neg B \Longleftrightarrow \neg(A \wedge B)$.
(3) $\neg A \wedge \neg B \Longleftrightarrow \neg(A \vee B)$.
(4) $A \to B \Longleftrightarrow \neg A \vee B$.

Proof. Each of these statements can be proved using a truth table, so from one example the reader may see how to do the others. Notice that truth tables give an algorithmic approach to questions of logical equivalence. The following truth table demonstrates equivalence (2) above.

	A	B	$(\neg A)$	$(\neg B)$	$((\neg A) \vee (\neg B))$	$(A \wedge B)$	$(\neg(A \wedge B))$
I_0	0	0	1	1	1	0	1
I_1	1	0	0	1	1	0	1
I_2	0	1	1	0	1	0	1
I_3	1	1	0	0	0	1	0
					$\uparrow$		$\uparrow$

$\square$

We want to have shorter expressions for some of the other important propositional sentences. In particular, let $A \leftrightarrow B$ denote $(A \wedge B) \vee (\neg A \wedge \neg B)$, which is true for the interpretations I with $I(A) = I(B)$. Let $A + B$ denote $\neg(A \leftrightarrow B)$, which is true in the interpretations where $I(A) \neq I(B)$. Let $\bot$ denote $A \wedge \neg A$, for some propositional variable A, which is false under any interpretation, and let T denote $A \vee \neg A$, which is true under any interpretation. Note that for any sentences A and B, $A \wedge \neg A \leftrightarrow B \wedge \neg B$, and likewise $A \vee \neg A \leftrightarrow B \vee \neg B$.

It can be seen from the observations above that every truth function of two propositional variables may be given by a sentence which uses only $\neg$ and $\vee$. It follows that every sentence is logically equivalent to a sentence constructed from $\neg$ and $\vee$. The following list gives three pairs of connectives each of which is sufficient to express all sentences. We will see in the next section that any truth function may be represented by a sentence.

$$\neg, \vee$$
$$\neg, \wedge$$
$$\neg, \rightarrow$$

We are going to study some notions about sets of sentences, such as *satisfiability*.

Definition 2.2.6. An *$\mathcal{L}$-theory* is a set of sentences in a language $\mathcal{L}$.

Definition 2.2.7. Fix a propositional language $\mathcal{L}$ and let Γ be an $\mathcal{L}$-theory.

(a) Γ *logically implies* a sentence P, in symbols, $\Gamma \models P$ if, for every interpretation I, if $I(Q) = 1$ for all $Q \in \Gamma$, then $I(P) = 1$.

(b) Γ is *satisfiable* if there is some interpretation I with $I(P) = 1$ for all $P \in \Gamma$.

(c) Γ is *complete* if, for every sentence P, either $P \in \Gamma$ or $\neg P \in \Gamma$.

Example 2.2.8. $\{(A \wedge B), (\neg C)\} \models (A \vee B)$. This is easy to verify using truth tables, or simply by observing that, whenever the hypotheses are true, then both A and B are true, and hence $A \vee B$ must be true.

The important *Satisfiability Problem* is to determine whether a given set of propositional sentence is satisfiable. This can be solved by means of truth tables. If a set of sentences has a total length of n, then it contains at most n propositional variables. Thus, the satisfiability may be checked by means of a truth table with 2^n rows. It is easy to check whether a given row makes the sentence true, so the truth table algorithm will take on the order of 2^{cn} steps; thus we say that the problem may be solved in exponential time. The famous $P = NP$ problem asks whether there is an algorithm which runs in polynomial time n^c. This was one of the seven unsolved Millennium Problems set forth by the Clay Mathematics Institute.

Example 2.2.9.

(1) The set $\{A \wedge \neg B, A \to B\}$ is not satisfiable. Suppose that

$$I(A \wedge \neg B) = 1 = I(A \to B).$$

Then $A \wedge \neg B$ is true, so that A is true and B is false. Also, $A \to B$ is true. Thus B is true. This contradiction shows that no truth interpretation can satisfy both sentences.

(2) The set $\{A \vee B, B \to C, \neg C\}$ is satisfiable by making A true, B false, and C false.

Exercises for Section 2.2

Exercise 2.2.1. Construct a truth table for the sentence $(A \lor B) \land (A \to \neg C)$.

Exercise 2.2.2. Use truth tables to show that $A \to (B \to C)$ and $(A \to B) \to C$ are not logically equivalent.

Exercise 2.2.3. Use truth tables to show that $\{(A \to B), (B \to C)\} \models (A \to C)$.

Exercise 2.2.4. Use truth tables to show that $\{(A \lor B), (B \lor C)\}$ does not logically imply $(A \lor C)$.

Exercise 2.2.5. Use truth tables to show that $A \to B \iff \neg A \lor B$.

Exercise 2.2.6. Use truth tables to show that $A \leftrightarrow B \iff (A \to B) \land (B \to A)$.

Exercise 2.2.7. Show that if $I(A)$ and $I(B)$ are defined, then $I(A \leftrightarrow B) = 1$.

Exercise 2.2.8. Use truth tables to check whether the following pairs of sentences are logically equivalent:

(a) $(A \lor B) \to C, (A \to C) \land (B \to C)$;
(b) $A \to (B \to C), (A \to B) \to C$.

Exercise 2.2.9. Investigate the following sets of formulas for satisfiability. For those that are satisfiable, give an interpretation which makes them all true. For those that are not satisfiable, give a brief explanation.

(a) $\{A \to (\neg(B \land C)), (D \lor E) \to G, G \to \neg(H \lor I), \neg C \land E \land H\}$.
(b) $\{A \to B, C \to D, B \to D, \neg C \to A, E \to G, G \to \neg D\}$.
(c) $\{(A \lor B) \to (C \land D), (D \lor E) \to G, A \lor \neg G\}$.
(d) $\{(A \to B) \land C, (D \to B) \land E, G \to \neg A, H \to I, \neg(\neg C \to E)\}$.

2.3. The Disjunctive Normal Form Theorem

In this section, we will show that the language of propositional calculus is sufficient to represent every possible truth function.

Definition 2.3.1.

(1) A *literal* is either a propositional variable or its negation.
(2) A *conjunctive clause* is a conjunction of literals.
(3) A *disjunctive clause* is a disjunction of literals.
(4) A propositional sentence is in *disjunctive form* if it is a disjunction of conjunctive clauses and it is in *conjunctive form* if it is a conjunction of disjunctive clauses.

A conjunctive clause C is trivial if it contains both A and $\neg A$ for some propositional variable A. We will generally assume that we are working with nontrivial conjunctive clauses. Note that any nontrivial conjunctive clause C is satisfiable, for example by making every literal appearing in C true. Now a disjunction $C_1 \vee C_2 \vee \cdots \vee C_n$ is satisfiable if and only if at least one of the conjunctive clauses C_i is satisfiable. Thus a sentence in disjunctive form is satisfiable whenever at least one of its conjunctive clauses is nontrivial. This means that the process of finding a disjunctive form which is logically equivalent to a given sentence P will also solve the satisfiability problem for P. The satisfiability problem is usually applied to sentences in conjunctive form. We write $\phi(A_1, \ldots, A_n)$ to indicate that a formula ϕ contains exactly the propositional variables $A_1, \ldots, A_n$. The following lemma is needed to prove the Disjunctive Normal Form Theorem.

Lemma 2.3.2.

(i) *For any nontrivial conjunctive clause $C = \phi(A_1, \ldots, A_n)$, there is a unique interpretation $I_C : \{A_1, \ldots, A_n\} \to \{0, 1\}$ such that $I_C(\phi) = 1$.*

(ii) *Conversely, for any interpretation* $I : \{A_1, \ldots, A_n\} \to \{0, 1\}$, *there is a unique conjunctive clause* C_I *(up to permutation of literals) such that* $I(C_I) = 1$ *and for any interpretation* $J \neq I$, $J(C_I) = 0$.

Proof.   (i) Let

$$B_i = \begin{cases} A_i & \text{if } C \text{ contains } A_i \text{ as a conjunct,} \\ \neg A_i & \text{if } C \text{ contains } \neg A_i \text{ as a conjunct.} \end{cases}$$

It follows that $C \iff B_1 \wedge \cdots \wedge B_n$. Now let $I_C(A_i) = 1$ if and only if $A_i = B_i$. Then clearly $I_C(B_i) = 1$ for $i = 1, 2, \ldots, n$ and therefore $I_C(C) = 1$. To show uniqueness, if $J(C) = 1$ for some interpretation J, then $\phi(B_i) = 1$ for each i and hence $J = I_C$.

(ii) Let

$$B_i = \begin{cases} A_i & \text{if } I(A_i) = 1, \\ \neg A_i & \text{if } I(A_i) = 0. \end{cases}$$

Let $C_I = B_1 \wedge \cdots \wedge B_n$. As above $I(C_I) = 1$ and $J(C_I) = 1$ imply that $J = I$.

It follows as above that I is the unique interpretation under which C_I is true. We claim that C_I is the unique conjunctive clause with this property. Suppose not. Then there is some conjunctive clause C' such that $I(C') = 1$ and $C' \neq C_I$. This implies that there is some propositional variable A in C' and $\neg A$ in C_I (or vice versa). But $I(C') = 1$ implies that $I(A) = 1$ and $I(C_I) = 1$ implies that $I(\neg A) = 1$, which is clearly impossible. Thus C_I is unique. $\square$

Here is the Disjunctive Normal Form Theorem.

Theorem 2.3.3. *For every* $n \in \mathbb{N}$, *and every function* $F : \{0, 1\}^n \to \{0, 1\}$, *there is a sentence* P *in disjunctive form such that* $F = TF_P$.

Proof. Let $I_1, I_2, \ldots, I_k$ be the interpretations in $\{0, 1\}^n$ such that $F(I_i) = 1$ for $i = 1, \ldots, k$. For each i, let $C_i = C_{I_i}$ be the conjunctive clauses guaranteed to hold by the previous lemma. Now let $P = C_1 \vee C_2 \vee \cdots \vee C_k$. Then for any interpretation I,

$$TF_P(I) = 1 \text{ if and only if } I(P) = 1 \text{ (by definition)}$$

$$\text{if and only if } I(C_i) = 1 \text{ for some } i = 1, \ldots, k$$

$$\text{if and only if } I = I_i \text{ for some } i \text{ (by the previous}$$

$$\text{lemma)}$$

$$\text{if and only if } F(I) = 1 \text{ (by the choice of } I_1, \ldots, I_k).$$

Hence $TF_P = F$ as desired. □

The sentence P constructed in the above theorem is said to be in *disjunctive normal form*, meaning that every conjunct contains each propositional variable.

Example 2.3.4. Suppose that we want a sentence $\phi(A_1, A_2, A_3)$ such that $I(\phi) = 1$ exactly for the three interpretations $(0, 1, 0)$, $(1, 1, 0)$ and $(1, 1, 1)$. The following formula ϕ will work, and it is in disjunctive normal form.

$$\phi = (\neg A_1 \wedge A_2 \wedge \neg A_3) \vee (A_1 \wedge A_2 \wedge \neg A_3) \vee (A_1 \wedge A_2 \wedge A_3).$$

It follows that the connectives $\neg, \wedge, \vee$ are sufficient to express all truth functions.

By De Morgan's Laws (2) and (3) of Lemma 2.2.5), the sets $\{\neg, \vee\}$ and $\{\neg, \wedge\}$ are also sufficient.

Exercises for Section 2.3

Exercise 2.3.1. Construct a sentence φ so that TF_φ has the values listed in the table below. This truth table represents the *exclusive* or, since the desired sentence is true if and only if exactly one of $\{A, B\}$ is true.

	A	B	TF_φ
I_0	0	0	0
I_1	1	0	1
I_2	0	1	1
I_3	1	1	0

Exercise 2.3.2. A *majority sentence* $\phi(A, B, C)$ is true if at least two of A, B, C are true and thus is false if at least two of the three are false. Construct a propositional sentence with this property.

Exercise 2.3.3. Let $|$ be the binary connective defined so that $A \mid B$ is true if and only if not both A and B are true. Show that this connective suffices to define negation and the usual binary connectives (and hence any possible sentence).

Exercise 2.3.4. Explain why the binary connectives $\{\vee, \wedge\}$ are not sufficient to define all other binary connectives.

Hint: Find a property of the truth functions which can be represented using only $\vee$ and $\wedge$.

2.4.   The Deductive Calculus

One of the basic tasks of a mathematician is proving theorems. This section develops a system rules of inference for propositional languages. This system may be viewed as a form of natural deduction, since it is based primarily on the definitions of each connective. With it one formalizes the notion of proof. Then one can ask questions about what can be proved, what cannot be proved, and how the notion of proof is related to the notion of interpretations. There are many alternative proof systems, such as natural deduction, the Hilbert calculus, and the Gentzen calculus; we follow the natural deduction proof system as developed by Suppes [32] and others.

The basic relation in the propositional calculus is the relation *proves* between a theory Γ, i.e., a collection of sentences, and a sentence B. A more long-winded paraphrase of the relation "Γ proves B" is "there is a formal proof of B using whatever hypotheses are needed from Γ". This relation is denoted $X \vdash Y$, with the following abbreviations for special cases:

Formal Version:	$\Gamma \vdash \{B\}$	$\{A\} \vdash B$	$\emptyset \vdash B$
Abbreviation:	$\Gamma \vdash B$	$A \vdash B$	$\vdash B$

We will first define the relation $\Gamma \vdash A$ between sets Γ of sentences and individual sentences A and then we will present the notion of a formal proof in the propositional calculus. The symbol $\vdash$ is sometimes called a (*single*) *turnstile*.

Definition 2.4.1. The relation $\Gamma \vdash B$ is the smallest subset of pairs (Γ, B) from $\mathcal{P}(\text{Sent}) \times \text{Sent}$ which is closed under the following rules of deduction.

(1) (Given Rule) If $B \in \Gamma$, then $\Gamma \vdash B$.

(2) ($\wedge$-Application) If $\Gamma \vdash (A \wedge B)$, then $\Gamma \vdash A$ and $\Gamma \vdash B$.

(3) ($\vee$-Application) If $\Gamma \vdash (A \vee B)$, $\Gamma \cup \{A\} \vdash C$, and $\Gamma \cup \{B\} \vdash C$, then $\Gamma \vdash C$.

(4) ($\rightarrow$-Application) If $\Gamma \vdash A \rightarrow B$ and $\Gamma \vdash A$, then $\Gamma \vdash B$.

(5) ($\bot$-Application) If $\Gamma \vdash \bot$, then $\Gamma \vdash A$ for any sentence A.

(6) ($\neg$-Application) If $\Gamma \vdash \neg\neg A$, then $\Gamma \vdash A$.

(7) ($\wedge$-Introduction) If $\Gamma \vdash A$ and $\Gamma \vdash B$, then $\Gamma \vdash A \wedge B$.

(8) ($\vee$-Introduction) If $\Gamma \vdash A$, then for any sentence B, $\Gamma \vdash A \vee B$ and $\Gamma \vdash B \vee A$.

(9) ($\rightarrow$-Introduction) If $\Gamma \cup \{A\} \vdash B$, then $\Gamma \vdash A \rightarrow B$.

(10) ($\bot$-Introduction) If $\Gamma \vdash A \wedge \neg A$, then $\Gamma \vdash \bot$.

(11) ($\neg$-Introduction) If $\Gamma \cup \{A\} \vdash \bot$, then $\Gamma \vdash \neg A$.

Note that in this definition the sentences A, B, C are variables representing arbitrary sentences and the set Γ likewise represents an arbitrary theory. The $\wedge$-Application rule is also known as "$\wedge$-Elimination" and similarly for rules 3, 4, 5, and 6. The $\vee$-Application rule is also known as "proof by cases". The $\rightarrow$-Application rule is the classic "modus ponens". Proofs using $\neg$-Introduction are often called proofs by contradiction.

It is immediate from $\wedge$-Introduction that $A \vdash A \wedge A$, and it is immediate from $\wedge$-Application that $A \wedge A \vdash A$. Similarly $A \vdash A \vee A$ follows from $\vee$-Introduction and $A \vee A \vdash A$ follows from $\vee$-Application.

Example 2.4.2. $A \wedge B \vdash B \wedge A$. We will demonstrate this in three ways, first, directly using Definition 2.4.1, second, writing out steps in sentence form, and third, by means of a formal proof.

Using Definition 2.4.1: First, $\{A \wedge B\} \vdash A \wedge B$ by Rule 1 (Given). Next, $\{A \wedge B\} \vdash A$ and $\{A \wedge B\} \vdash B$ by Rule 2. Then $\{A \wedge B\} \vdash B \wedge A$ by Rule 7 ($\wedge$-Introduction).

In sentence form: $A \wedge B$ is given. Thus we have both A and B by $\wedge$-Application. Since we have both A and B, we can conclude $B \wedge A$ by $\wedge$-Introduction.

A formal proof:

1	$A \wedge B$	Given
2	A	$\wedge$-Application 1
3	B	$\wedge$-Application 1
4	$B \wedge A$	$\wedge$-Introduction 2–3.

It is important to observe that in several of the rules we use a set Γ as well as the set $\Gamma \cup \{A\}$ and $\Gamma \cup \{B\}$. When writing proofs, this means that we will make temporary assumptions as part of the argument, and then later remove those assumptions.

Example 2.4.3. Let us show that $\vdash A \to A$, that is, we prove $A \to A$.

Using Definition 2.4.1: First, $\{A\} \vdash A$ by Rule 1 (Given). Thus $\emptyset \vdash A \to A$ by Rule 9 ($\to$-Introduction).

In sentence form: Let A be a temporary assumption. Then we may conclude A as given. Thus we have deduced A from assumption A. It follows from $\to$-Introduction that $A \to A$.

A formal proof:

$$
\begin{array}{lll}
1 & \quad A & \text{Assumption} \\
2 & \quad A & \text{Given} \\
3 & A \to A & \to\text{-Introduction 1–2.}
\end{array}
$$

Here the vertical bar indicates that these first two lines are subject to the temporary assumption of A, whereas the third line holds without any assumptions.

Here we make the definition of a formal proof a bit more precise. Observe that in the course of a proof, additional temporary assumptions are introduced by $\vee$-Application, $\to$-Introduction, and $\neg$-Introduction, and later removed.

Definition 2.4.4. A *formal proof* or derivation of a propositional sentence P from a collection of propositional sentences Γ is a finite sequence $P_1, P_2, \ldots, P_n$ of propositional sentences where P_n is the sentence P, together with a sequence $\Delta_1, \ldots, \Delta_n = \emptyset$ of temporary assumptions needed for each P_i, where each sentence in the sequence is either in Γ or is obtained from sentences occurring earlier in the sequence by means of one of the rules from Definition 2.4.1, with the corresponding modification of the set of temporary hypotheses required by the given rule.

As shown in Example 2.4.3, temporary hypotheses are indicated by an indentation and a remark.

We now provide some examples of proofs, as well as exercises.

Proposition 2.4.5. *For any sentences A, B, C,*

(1) $A \to B \vdash \neg B \to \neg A$;

(2) $\{A \to B, B \to C\} \vdash A \to C$;

(3) $\{A \vee B, \neg A\} \vdash B$;

(4) $A \vdash \neg\neg A$;

(5) (a) $A \vee B \vdash B \vee A$ *and* (b) $A \wedge B \vdash B \wedge A$;

(6) (a) $(A \vee B) \vee C \vdash A \vee (B \vee C)$ *and* (b) $A \vee (B \vee C) \vdash$ $(A \vee B) \vee C$;

(7) (a) $(A \wedge B) \wedge C \vdash A \wedge (B \wedge C)$ *and* (b) $A \wedge (B \wedge C) \vdash$ $(A \wedge B) \wedge C$;

(8) (a) $A \wedge (B \vee C) \vdash (A \wedge B) \vee (A \wedge C)$ *and* (b) $(A \wedge B) \vee$ $(A \wedge C) \vdash A \wedge (B \vee C)$;

(9) *(a)* $A \vee (B \wedge C) \vdash (A \vee B) \wedge (A \vee C)$ *and* (b) $(A \vee B) \wedge$ $(A \vee C) \vdash A \vee (B \wedge C)$;

(10) (a) $\neg(A \vee B) \vdash \neg A \wedge \neg B$ *and* (b) $\neg A \wedge \neg B \vdash \neg(A \vee B)$;

(11) $\vdash A \vee \neg A$;

(12) (a) $\neg(A \wedge B) \vdash \neg A \vee \neg B$ *and* (b) $\neg A \vee \neg B \vdash \neg(A \wedge B)$;

(13) (a) $\neg A \vee B \vdash A \to B$ *and* (b) $A \to B \vdash \neg A \vee B$.

We give brief sketches of some of these proofs to illustrate the various methods.

Proof.   (2) $\{A \to B, B \to C\} \vdash A \to C$

1	$A \to B$	Given
2	$B \to C$	Given
3	A	Assumption
4	B	$\to$-Application 1, 3
5	C	$\to$-Application 2, 4
6	$A \to C$	$\to$-Introduction 3–5.

(3) $\{A \lor B, \neg A\} \vdash B$

1	$A \lor B$	Given
2	$\neg A$	Given
3	A	Assumption
4	$A \land \neg A$	$\land$-Introduction 2, 3
5	$\bot$	$\bot$-Introduction 4
6	B	$\bot$-Application 5
7	B	Assumption
8	B	Given
9	B	$\lor$-Application 1–8.

In sentence form: We are given $\neg A$ as well as $A \lor B$. There are two cases toward $\lor$-Application. First suppose A. This contradicts the assumption $\neg A$. Thus we may conclude B by $\bot$-Application. Second suppose B. In either case, we have B.

(4) $A \vdash \neg\neg A$

1	A	Given
2	$\neg A$	Assumption
3	$A \land \neg A$	$\land$-Introduction 1, 2
4	$\bot$	$\bot$-Introduction 3
5	$\neg\neg A$	$\neg$-Introduction 1–4.

(5) (a) $A \vee B \vdash B \vee A$

1	$A \vee B$	Given
2	A	Assumption
3	$B \vee A$	$\vee$-Introduction 2
4	B	Assumption
5	$B \vee A$	$\vee$-Introduction 4
6	$B \vee A$	$\vee$-Application 1–5.

See Example 2.4.2 above for part (b).

(7) (a) $(A \wedge B) \wedge C \vdash A \wedge (B \wedge C)$

1	$(A \wedge B) \wedge C$	Given
2	$A \wedge B$	$\wedge$-Application 1
3	A	$\wedge$-Application 2
4	B	$\wedge$-Application 2
5	C	$\wedge$-Application 1
6	$B \wedge C$	$\wedge$-Introduction 4, 5
7	$A \wedge (B \wedge C)$	$\wedge$-Introduction 3, 6.

(10) (a) $\neg(A \lor B) \vdash \neg A \land \neg B$

1	$\neg(A \lor B)$	Given
2	A	Assumption
3	$A \lor B$	$\lor$- Introduction 2
4	$(A \lor B) \land \neg(A \lor B)$	$\land$-Introduction 1, 3
5	$\bot$	$\bot$-Introduction 4
6	$\neg A$	$\neg$-Introduction 2–5
7	B	Assumption
8	$A \lor B$	$\lor$- Introduction 7
9	$(A \lor B) \land \neg(A \lor B)$	$\land$-Introduction 1, 8
10	$\bot$	$\bot$-Introduction 9
11	$\neg B$	$\neg$-Introduction 7–10
12	$\neg A \land \neg B$	$\land$-Introduction 6, 11.

In sentence form: $\neg(A \lor B)$ is given. Assume A by way of contradiction. Then we get $A \lor B$ by $\lor$-Introduction. This contradicts the given. Thus we may deduce $\neg A$ by $\neg$-Introduction. Similarly, the assumption of B leads to a contradiction, giving us $\neg B$. By $\land$-Introduction, we obtain $\neg A \land \neg B$, as desired.

(11) $\vdash A \lor \neg A$

1	$\neg(A \lor \neg A)$	Assumption
2	$\neg A \land \neg\neg A$	Item 10(a), 1
3	$\bot$	$\bot$-Introduction 2
4	$\neg\neg(A \lor \neg A)$	$\neg$-Introduction 1–3
5	$A \lor \neg A$	$\neg$-Application.

(12) (b) $\neg A \lor \neg B \vdash \neg(A \land B)$

1	$\neg A \lor \neg B$	Given
2	$A \land B$	Assumption
3	A	$\land$-Application 2
4	$\neg\neg A$	Item 6, 3
5	$\neg B$	Item 4, 1, 4
6	B	$\land$-Application 2
7	$B \land \neg B$	$\land$-Introduction 5, 6
8	$\bot$	$\bot$-Introduction 7
9	$\neg(A \land B)$	$\neg$ -Introduction 2–8.

(13) (a) $\neg A \lor B \vdash A \to B$

1	$\neg A \lor B$	Given
2	A	Assumption
3	$\neg\neg A$	Item 6, 2
4	B	Item 4, 1,3
5	$A \to B$	$\to$-Introduction 2–4.

(b) $A \to B \vdash \neg A \vee B$

1	$A \to B$	Given
2	$\neg(\neg A \vee B)$	Assumption
3	$\neg\neg A \wedge \neg B$	Item 12, 2
4	$\neg\neg A$	$\wedge$-Application 3
5	A	$\neg$-Application 4
6	B	$\to$-Application 1, 5
7	$\neg B$	$\wedge$-Application 3
8	$B \wedge \neg B$	$\wedge$-Application 6, 7
9	$\bot$	$\bot$-Introduction 8
10	$\neg\neg(\neg A \vee B)$	$\neg$-Introduction 2-9
11	$\neg A \vee B$	$\neg$-Application 10 $\qquad\square$

The following general properties about $\vdash$ will be useful when we prove the soundness and completeness theorems.

Lemma 2.4.6 (Deduction Lemma). *For any theory Γ and any sentences A and B, if $\Gamma \vdash A$ and $\Gamma \cup \{A\} \vdash B$, then $\Gamma \vdash B$.*

Proof. $\Gamma \cup \{A\} \vdash B$ implies $\Gamma \vdash A \to B$ by $\to$-Introduction. Combining this latter fact with the fact that $\Gamma \vdash A$ yields $\Gamma \vdash B$ by $\to$-Application. $\qquad\square$

Proposition 2.4.7. *For any theories Γ and Δ and any sentence B, if $\Gamma \vdash B$ and $\Gamma \subseteq \Delta$, then $\Delta \vdash B$.*

Proof. This follows by induction on proof length. For the base case, if B follows from Γ on the basis of the Given Rule, then it must be the case that $B \in \Gamma$. Since $\Gamma \subseteq \Delta$ it follows that $B \in \Delta$ and hence $\Delta \vdash B$ by the Given Rule.

If the final step in the proof of B from Γ is made on the basis of any one of the rules, then we may assume by the induction hypothesis that the other formulas used in these deductions follow from Δ (since they follow from Γ). We will look at two cases and leave the rest to the reader.

Suppose that the last step comes by $\rightarrow$-Application, where we have derived $A \rightarrow B$ and A from Γ earlier in the proof. Then we have $\Gamma \vdash A \rightarrow B$ and $\Gamma \vdash B$. By the induction hypothesis, $\Delta \vdash A$ and $\Delta \vdash A \rightarrow B$. Hence $\Delta \vdash B$ by $\rightarrow$-Elimination.

Suppose that the last step comes from $\wedge$-Application, where we have derived $A \wedge B$ from Γ earlier in the proof. Since $\Gamma \vdash A \wedge B$, by inductive hypothesis it follows that $\Delta \vdash A \wedge B$. Hence $\Delta \vdash B$ by $\wedge$-Application. $\qquad \square$

Here is a version of *compactness* for the notion of proof.

Theorem 2.4.8. *For any theory Γ, if $\Gamma \vdash B$, then there is a finite set $\Gamma_0 \subseteq \Gamma$ such that $\Gamma_0 \vdash B$.*

Proof. Again we argue by induction on the length of the proof of $\Gamma \vdash B$. For the base case, if B follows from Γ on the basis of the Given Rule, then $B \in \Gamma$ and we can let $\Gamma_0 = \{B\}$.

If the final step in the proof of B from Γ is made on the basis of any one of the rules, then we may assume by the induction hypothesis that the other formulas used in these deductions follow from some finite $\Gamma_0 \subseteq \Gamma$. We will look at two cases and leave the rest to the reader.

Suppose that the last step of the proof comes by $\vee$-Introduction, so that B is of the form $C \vee D$. Then, without loss of generality, we can assume that we derived C from Γ earlier in the proof. Thus $\Gamma \vdash C$. By the induction hypothesis, there is a finite $\Gamma_0 \subseteq \Gamma$ such that $\Gamma_0 \vdash C$. Hence by $\vee$-Introduction, $\Gamma_0 \vdash C \vee D$.

Suppose that the last step of the proof comes by ∨-Application. Then earlier in the proof

(i) we have derived some formula $C \vee D$ from Γ,
(ii) under the assumption of C we have derived B from Γ, and
(iii) under the assumption of D we have derived B from Γ.

Thus, $\Gamma \vdash C \vee D$, $\Gamma \cup \{C\} \vdash B$, and $\Gamma \cup \{D\} \vdash B$. Then by assumption, by the induction hypothesis, there exist finite sets Γ_0, Γ_1, and Γ_2 of Γ such that $\Gamma_0 \vdash C \vee D$, $\Gamma_1 \cup \{C\} \vdash B$ and $\Gamma_2 \cup \{D\} \vdash B$. By Proposition 2.4.7,

(i) $\Gamma_0 \cup \Gamma_1 \cup \Gamma_2 \vdash C \vee D$,
(ii) $\Gamma_0 \cup \Gamma_1 \cup \Gamma_2 \cup \{C\} \vdash B$,
(iii) $\Gamma_0 \cup \Gamma_1 \cup \Gamma_2 \cup \{D\} \vdash B$.

Thus by ∨-Application, we have $\Gamma_0 \cup \Gamma_1 \cup \Gamma_2 \vdash B$. Since $\Gamma_0 \cup \Gamma_1 \cup \Gamma_2$ is finite and $\Gamma_0 \cup \Gamma_1 \cup \Gamma_2 \subseteq \Gamma$, the result follows. □

Definition 2.4.9. A set Γ of sentences is *inconsistent* if Γ leads to a contradiction, that is, if $\Gamma \vdash \bot$. Otherwise Γ is *consistent*.

Example 2.4.10. The set $\{A, \neg A\}$ is inconsistent for any sentence A.

We will demonstrate in Section 2.6 that being consistent is equivalent to being satisfiable.

Proposition 2.4.11. *The following are equivalent for all theories Γ:*

(1) Γ *is inconsistent;*
(2) *for every sentence A, $\Gamma \vdash A$;*
(3) *there is some sentence A such that $\Gamma \vdash A$ and $\Gamma \vdash \neg A$.*

The proof is left as an exercise.

Exercises for Section 2.4

Exercise 2.4.1. Give a proof that $A \to B \vdash \neg B \to \neg A$.

Exercise 2.4.2. Prove that $(A \wedge B) \to C \vdash (A \to (B \to C))$.

Exercise 2.4.3. Give a proof for (6(a)): $(A \vee B) \vee C \vdash A \vee (B \vee C)$.

Exercise 2.4.4. Give a proof for (6(b)): (b) $A \vee (B \vee C) \vdash (A \vee B) \vee C$.

Exercise 2.4.5. Give a proof for (7(b)): $A \wedge (B \wedge C) \vdash (A \wedge B) \wedge C$.

Exercise 2.4.6. Give a proof for (8(a)): $A \wedge (B \vee C) \vdash (A \wedge B) \vee (A \wedge C)$.

Exercise 2.4.7. Give a proof for (8(b)): $(A \wedge B) \vee (A \wedge C) \vdash A \wedge (B \vee C)$.

Exercise 2.4.8. Give a proof for (9(a)): $A \vee (B \wedge C) \vdash (A \vee B) \wedge (A \vee C)$.

Exercise 2.4.9. Give a proof for (9(b)): $(A \vee B) \wedge (A \vee C) \vdash A \vee (B \wedge C)$.

Exercise 2.4.10. Give a proof for (10(b)): $\neg A \wedge \neg B \vdash \neg (A \vee B)$.

Exercise 2.4.11. Give a proof for (12(a)): $\neg (A \wedge B) \vdash \neg A \vee \neg B$.

Exercise 2.4.12. Give a proof in sentence form for (12(b)): $\neg (A \wedge B) \vdash \neg A \vee \neg B$.

Exercise 2.4.13. Give a proof in sentence form for (13(b)): $A \to B \vdash \neg A \vee B$.

Exercise 2.4.14. Give the case for $\neg$-Application in the proof of Theorem 2.4.8.

Exercise 2.4.15. Give the case for $\to$-Application in the proof of Theorem 2.4.8.

Exercise 2.4.16. Give the case for ¬-Introduction in the proof of Theorem 2.4.8.

Exercise 2.4.17. Give the case for →-Introduction in the proof of Theorem 2.4.8.

Exercise 2.4.18. Show that a theory Γ is consistent if and only if there is some sentence A such that $\Gamma \nvdash A$.

Exercise 2.4.19. Show that a theory Γ is inconsistent if and only if there is some sentence A such that $\Gamma \vdash A$ and $\Gamma \vdash \neg A$.

2.5. The Soundness Theorem

We now determine the precise relationship between $\vdash$ and $\models$ for propositional logic. Our first major theorem says that if one can prove something in A from a theory Γ, then Γ logically implies A.

Theorem 2.5.1 (Soundness Theorem, Version I). *For any theory Γ, if $\Gamma \vdash A$, then $\Gamma \models A$.*

Theorem 2.5.2 (Soundness Theorem, Version II). *For any theory Γ, if Γ is satisfiable, then Γ is consistent.*

We will show that Version I implies Version II and then prove Version I. Let us assume Version I of the Soundness Theorem and prove Version II. Suppose therefore that Γ is satisfiable and let I be an interpretation such that $I(P) = 1$ for all $P \in \Gamma$. Now suppose by way of contradiction that Γ is not consistent. Then $\Gamma \vdash \bot$. Thus, by Version I of the Soundness Theorem, $\Gamma \models \bot$. But this implies that $I(\bot) = 1$, which is not possible.

Now we give the proof of Version I.

Proof. The proof is by induction on the length n of the proof of A. We need to show that if there is a proof of A from Γ, then for any interpretation I such that $I(\gamma) = 1$ for all $\gamma \in \Gamma$, $I(A) = 1$.

Base Step: $n = 1$. For a one-step deduction, we must have used the Given Rule, so that $A \in \Gamma$. If the truth interpretation I has $I(\gamma) = 1$ for all $\gamma \in \Gamma$, then of course $I(A) = 1$ since $A \in \Gamma$.

Induction Step: $n > 0$. Assume the theorem holds for all shorter deductions. Now proceed by cases on the other rules. We prove a few examples and leave the rest for the reader.

Suppose that the last step of the deduction is given by $\vee$-Introduction, so that A has the form $B \vee C$. Without loss of generality, suppose we have derived B from Γ earlier in the proof. Suppose that $I(\gamma) = 1$ for all $\gamma \in \Gamma$. Since the proof of $\Gamma \vdash B$ is shorter than the given deduction of $B \vee C$, by the induction hypothesis, $I(B) = 1$. But then $I(B \vee C) = 1$ since I is an interpretation.

Suppose that the last step of the deduction is given by $\wedge$-Application. Suppose that $I(\gamma) = 1$ for all $\gamma \in \Gamma$. Without loss of generality A has been derived from a sentence of the form $A \wedge B$, which has been derived from Γ in a strictly shorter proof. Since $\Gamma \vdash A \wedge B$, it follows by induction hypothesis that $\Gamma \models A \wedge B$, and hence $I(A \wedge B) = 1$. Since I is an interpretation, it follows that $I(A) = 1$.

Suppose that the last step of the deduction is given by $\rightarrow$-Introduction. Then A has the form $B \rightarrow C$. It follows that under the assumption of B, we have derived C from Γ. Thus $\Gamma \cup \{B\} \vdash C$ in a strictly shorter proof. Suppose that $I(\gamma) = 1$ for all $\gamma \in \Gamma$. We have two cases to consider. First, if $I(B) = 0$, it follows that $I(B \rightarrow C) = 1$. Second, if $I(B) = 1$, then since $\Gamma \cup \{B\} \vdash C$, it follows that $I(C) = 1$. Then $I(B \rightarrow C) = 1$. In either case, the conclusion follows.

Suppose the last step of the deduction is by $\perp$-Application. That is, $\Gamma \vdash \perp$ via a proof which is one step shorter, and then $\Gamma \vdash A$ at the final step. Let I be a truth interpretation such that $I(\gamma) = 1$ for all $\gamma \in \Gamma$. Then by the induction hypothesis, $I(\perp) = 1$. However, $I(\perp) = 0$ for any truth interpretation. Therefore no such interpretation can exist. Thus it is vacuously true that for any truth interpretation I such that $I(\gamma) = 1$ for all $\gamma \in \Gamma$, $I(A) = 1$. □

In Section 2.6, we will prove the converse of the Soundness Theorem by showing that any consistent theory is satisfiable.

Exercises for Section 2.5

Exercise 2.5.1. Give the case for ¬-Introduction in the proof of Theorem 2.5.1.

Exercise 2.5.2. Give the case for ¬-Application in the proof of Theorem 2.5.1.

Exercise 2.5.3. Give the case for ∨-Application in the proof of Theorem 2.5.1.

2.6.   The Completeness Theorem

In this section, we will demonstrate the completeness of our proof system.

For the proof system, completeness means that the system is sufficient, or large enough, to prove every tautology. On the other hand, the notion of completeness for a theory, as defined in Section 2.2, is that a theory Γ is complete if, for every sentence P, either $P \in \Gamma$ or $\neg P \in \Gamma$. So a theory is complete if it is sufficient, or large enough, to determine every sentence. Thus we see that the word "complete" is used in two different senses.

Fix a propositional language $\mathcal{L}$. The Completeness Theorem for $\mathcal{L}$ demonstrates that the list of axioms and rules of propositional calculus are sufficient to prove all true sentences. We will omit any mention of $\mathcal{L}$ for the remainder of the section.

Theorem 2.6.1 (The Completeness Theorem, Version I).
For any theory Γ and any sentence A, if $\Gamma \models A$, then $\Gamma \vdash A$.

Theorem 2.6.2 (The Completeness Theorem, Version II).
If a theory Γ is consistent, then Γ is satisfiable.

We will show that Version II implies Version I and then prove Version II. First we give alternate versions of the Compactness Theorem (Theorem 2.6.3).

Theorem 2.6.3. *For any theory Δ, if every finite subset of Δ is consistent, then Δ is consistent.*

Proof. We show the contrapositive. Suppose that Δ is not consistent. Then, for some B, $\Delta \vdash B \wedge \neg B$. It follows from Theorem 2.4.8 that Δ has a finite subset Δ_0 such that $\Delta_0 \vdash B \wedge \neg B$. But then Δ_0 is not consistent. $\square$

Theorem 2.6.4. *Let Δ and $\Delta_0, \Delta_1, \Delta_2, \ldots$ be theories and suppose that*

(i) $\Delta = \bigcup_n \Delta_n$,
(ii) $\Delta_n \subseteq \Delta_{n+1}$ *for every n, and*
(iii) Δ_n *is consistent for each n.*

Then Δ is consistent.

Proof. Again we show the contrapositive. Suppose that Δ is not consistent. Then by Theorem 2.6.3, Δ has a finite, inconsistent subset $F = \{\delta_1, \delta_2, \ldots, \delta_k\}$. Since $\Delta = \bigcup_n \Delta_n$, there exists, for each $i \leq k$, some n_i such that $\delta_i \in \Delta_{n_i}$. Letting $n = \max\{n_i : i \leq k\}$, it follows that $F \subseteq \Delta_n$. But then Δ_n is inconsistent. $\square$

Several lemmas are needed in the proof of the Completeness Theorem.

Lemma 2.6.5. *For any theory Γ and any sentence A, $\Gamma \vdash A$ if and only if $\Gamma \cup \{\neg A\}$ is inconsistent.*

Proof. Suppose first that $\Gamma \vdash A$. Then $\Gamma \cup \{\neg A\}$ proves both A and $\neg A$ and is therefore inconsistent.

Suppose next that $\Gamma \cup \{\neg A\}$ is inconsistent. It follows from ¬-Introduction that $\Gamma \vdash \neg\neg A$. Then by ¬-Application, $\Gamma \vdash A$.

$\square$

We are now in position to show that Version II of the Completeness Theorem implies Version I. We show the contrapositive of the statement of Version I; that is, we show $\Gamma \not\vdash A$ implies $\Gamma \not\models A$. Suppose it is not the case that $\Gamma \vdash A$. Then by Lemma 2.6.5, $\Gamma \cup \{\neg A\}$ is consistent. Thus by Version II, $\Gamma \cup \{\neg A\}$ is satisfiable. Then it is not the case that $\Gamma \models A$.

Lemma 2.6.6. *If a theory Γ is consistent, then for any A, either $\Gamma \cup \{A\}$ is consistent or $\Gamma \cup \{\neg A\}$ is consistent.*

Proof. Suppose that $\Gamma \cup \{\neg A\}$ is inconsistent. Then by the previous lemma, $\Gamma \vdash A$. Then, for any B, it follows from the Deduction Lemma (Lemma 2.4.6) that $\Gamma \cup \{A\} \vdash B$ if and only if $\Gamma \vdash B$. Since Γ is consistent, it follows that $\Gamma \cup \{A\}$ is also consistent. $\square$

Lemma 2.6.7. *Suppose that a theory Δ is consistent and complete.*

(1) *For any sentence A, $\neg A \in \Delta$ if and only if $A \notin \Delta$.*
(2) *For any sentence A, if $\Delta \vdash A$, then $A \in \Delta$.*

Proof. (1) If $\neg A \in \Delta$, then $A \notin \Delta$ since Δ is consistent. If $A \notin \Delta$, then $\neg A \in \Delta$ since Δ is complete.

(2) Suppose that $\Delta \vdash A$ and suppose by way of contradiction that $A \notin \Delta$. Then by part (1), $\neg A \in \Delta$. But this contradicts the consistency of Δ. $\square$

Proposition 2.6.8. *A consistent theory Γ is complete if and only if it is* maximally consistent, *that is, no proper extension of Γ is consistent.*

The proof is left as an exercise.

Proposition 2.6.9. *For any complete, consistent theory Δ, there is a unique truth interpretation I which is satisfied by Δ.*

Proof.   Let Δ be a complete, consistent theory and define the function $I : Sent \to \{0, 1\}$ as follows. For each sentence B,

$$I(B) = \begin{cases} 1 & \text{if } B \in \Delta, \\ 0 & \text{if } B \notin \Delta. \end{cases}$$

We will show that I is a truth interpretation and $I(B) = 1$ for all $B \in \Delta$. First, we show that I preserves the four connectives: $\neg, \vee, \wedge$, and $\to$. We will show the first two and leave the others as exercises.

($\neg$): It follows from the definition of I and Lemma 2.6.7 that $I(\neg A) = 1$ if and only if $\neg A \in \Delta$ if and only if $A \notin \Delta$ if and only if $I(A) = 0$.

($\vee$): Suppose that $I(A \vee B) = 1$. Then $A \vee B \in \Delta$. We argue by cases. If $A \in \Delta$, then clearly $\max\{I(A), I(B)\} = 1$. Now suppose that $A \notin \Delta$. Then by completeness, $\neg A \in \Delta$. It follows from Proposition 2.4.5(3) that $\Delta \vdash B$. Hence $B \in \Delta$ by Lemma 2.6.7. Thus $\max\{I(A), I(B)\} = 1$.

Next suppose that $\max\{I(A), I(B)\} = 1$. Without loss of generality, $I(A) = 1$ and hence $A \in \Delta$. Then $\Delta \vdash A \vee B$ by $\vee$-Introduction, so that $A \vee B \in \Delta$ by Lemma 2.6.7 and hence $I(A \vee B) = 1$.

To establish the uniqueness of I, suppose that J is any interpretation satisfied by Δ. If $B \in \Delta$, then $J(B) = 1 = I(B)$. On the other hand, if $B \notin \Delta$, then $\neg B \in \Delta$, so that $J(\neg B) = 1$ and thus $J(B) = 0 = I(B)$.   □

We now prove Version II of the Completeness Theorem.

Proof of Theorem 2.6.2.   Let Γ be a consistent set of propositional sentences. Let $A_0, A_1, \dots$ be an enumeration of the set of sentences. We will define a sequence $\Delta_0 \subseteq \Delta_1 \subseteq \cdots$ and let

$\Delta = \cup_n \Delta_n$. We will show that Δ is a complete and consistent extension of Γ and then define an interpretation $I = I_\Delta$ to show that Γ is satisfiable.

$\Delta_0 = \Gamma$ and, for each n,

$$\Delta_{n+1} = \begin{cases} \Delta_n \cup \{A_n\}, & \text{if } \Delta_n \cup \{A_n\} \text{ is consistent,} \\ \Delta_n \cup \{\neg A_n\}, & \text{otherwise.} \end{cases}$$

It follows from the construction that, for each sentence A_n, either $A_n \in \Delta_{n+1}$ or $\neg A_n \in \Delta_{n+1}$. Hence Δ is complete. It remains to show that Δ is consistent.

Claim 1: *For each n, Δ_n is consistent.*

Proof of Claim 1: The proof is by induction on n. For the base case, we are given that $\Delta_0 = \Gamma$ is consistent. For the induction step, suppose that Δ_n is consistent. Then by Lemma 2.6.6, either $\Delta_n \cup \{A_n\}$ is consistent, or $\Delta_n \cup \{\neg A_n\}$ is consistent. In the first case $\Delta_{n+1} = \Delta_n \cup \{A_n\}$ and hence Δ_{n+1} is consistent. In the second case, $\Delta_{n+1} = \Delta_n \cup \{\neg A_n\}$ and hence Δ_{n+1} is consistent.

Claim 2: Δ *is consistent.*

Proof of Claim 2: This follows immediately from Theorem 2.6.4.

It now follows from Proposition 2.6.9 that there is a truth interpretation I such that $I(Q) = 1$ for all $Q \in \Delta$. Since $\Gamma \subseteq \Delta$, this proves that Γ is satisfiable. □

Here is an illustration of the construction above. Suppose that

$$\Gamma = \{A_0 \vee A_1 \vee A_2, \neg A_1 \vee \neg A_2 \vee \neg A_3, A_2 \vee A_3 \vee A_4,$$

$$\neg A_3 \vee \neg A_4 \vee \neg A_5, \dots \}.$$

Then $\Delta = \{A_0, A_1, A_2, \neg A_3, A_4, A_5, A_6, \neg A_7, \dots\}$ will result from the construction. We put A_0, A_1, A_2 into Δ_3 since it is consistent to do so, but after that the sentence $\neg A_1 \vee \neg A_2 \vee \neg A_3$ now makes adding A_3 result in an inconsistency.

Exercises for Section 2.6

Exercise 2.6.1. Give the case for the connective $\wedge$ in the proof of Proposition 2.6.9.

Exercise 2.6.2. Give the case for the connective $\rightarrow$ in the proof of Proposition 2.6.9.

Exercise 2.6.3. Show that if new propositional variables are added to a language $\mathcal{L}$, this does not change the provability of sentences in $\mathcal{L}$.

Exercise 2.6.4. Show that for any consistent theory Γ, Γ is maximally consistent if and only if Γ is complete.

2.7. Completeness, Consistency, and Independence

In this section, we consider the notions of completeness, consistency, and independence of a theory, as well as the related notion of axiomatizability.

Definition 2.7.1. Let Γ be a theory in a propositional language $\mathcal{L}$.

(a) The deductive closure $\text{DC}(\Gamma)$ is the set $\{B : \Gamma \vdash B\}$. This is the set of logical consequences of Γ.
(b) Γ is a *deductive theory* if it is closed under deduction.
(c) Γ is *independent* if Γ has no proper subset Δ such that $\text{DC}(\Delta) = \text{DC}(\Gamma)$; this means that Γ is minimal among the sets Δ with $\text{DC}(\Delta) = \text{DC}(\Gamma)$.

It is easy to see that Γ is a deductive theory if and only if $\text{DC}(\Gamma) = \Gamma$. This is left as an exercise. Then we observe that Γ is consistent if there is no sentence B such that $B \in \text{DC}(\Gamma)$ and $\neg B \in \text{DC}(\Gamma)$. Furthermore, $\text{DC}(\Gamma)$ is complete if for every sentence B, either $\Gamma \vdash B$ or $\Gamma \vdash \neg B$. For example, in the language $\mathcal{L}$ with two propositional variables A, B, the set $\{A, \neg B\}$

is independent and has a complete theory. This set is equivalent to the single sentence $A \wedge \neg B$. In the language $\mathcal{L}$ with three propositional variables A, B, C, the set $\{A, B, C\}$ is independent and its theory is complete.

Definition 2.7.2. A set Γ of sentences is said to be *finitely axiomatizable* if there is a finite set Δ such that $\mathrm{DC}(\Delta) = \mathrm{DC}(\Gamma)$.

For example, let $\Gamma = \{Q_n : n \in \mathbb{N}\}$, where $Q_0 = A$, and let $Q_{n+1} = A \wedge Q_n$ for each n, so that Q_n is the conjunction of n copies of A. Then clearly each Q_n is logically equivalent to A, so that $\mathrm{DC}(\Gamma) = \mathrm{DC}(\{A\})$.

Proposition 2.7.3. *For any theory* Γ, Γ *is finitely axiomatizable if and only if there is a single sentence A such that* $\mathrm{DC}(\Gamma) = \mathrm{DC}(A)$.

Proof. Suppose that Γ is finitely axiomatizable and let $\mathrm{DC}(\Gamma) = \mathrm{DC}(\{A_1, \ldots, A_n\})$. Then in turn $\mathrm{DC}(\{A_1, \ldots, A_n\}) = \mathrm{DC}(\{A_1 \wedge \cdots \wedge A_n\})$.

Proposition 2.7.4. *If a theory* Γ *is infinite and independent, then it is not finitely axiomatizable.*

Proof. Let $\Gamma = \{Q_0, Q_1, \ldots\}$ be independent and suppose by way of contradiction that $\mathrm{DC}(\Gamma) = \mathrm{DC}(\{A\})$ for some sentence A. It follows that $\Gamma \vdash A$ and also that $A \vdash Q_i$ for each i. Since $\Gamma \vdash A$, it follows by compactness that $\{Q_0, \ldots, Q_n\} \vdash A$ for some n. But this means that $\{Q_0, \ldots, Q_n\} \vdash Q_i$ for all i, so that $\mathrm{DC}(\{Q_0, \ldots, Q_n\}) = \mathrm{DC}(\Gamma)$, contradicting the independence of Γ. $\qquad\square$

The following propositions characterize independence as well as being simultaneously complete and consistent.

Proposition 2.7.5. *A theory* Γ *is independent if and only if for every $B \in \Gamma$, it is not the case that $\Gamma \setminus \{B\} \vdash B$.*

Proof. We prove the contrapositive. Suppose first that Γ is not independent. Then there is a proper subset Δ of Γ such that $\mathrm{DC}(\Delta) = \mathrm{DC}(\Gamma)$. Let $B \in \Gamma \setminus \Delta$, so that $\Delta \subseteq \Gamma \setminus \{B\}$. It follows that $\Gamma \setminus \{B\} \vdash B$. For the other direction, suppose that there exists B such that $\Gamma \setminus \{B\} \vdash B$. Then $\mathrm{DC}(\Gamma \setminus \{B\}) = \mathrm{DC}(\Gamma)$.

$\square$

Proposition 2.7.6. *A deductive theory Γ is complete and consistent if and only if there is a unique interpretation I satisfied by Γ.*

Proof. Suppose that Γ is complete and consistent. Then Γ is maximally consistent and has the unique truth interpretation given by Proposition 2.6.9, so that $I(B) = 1$ if and only if $B \in \Gamma$.

For the other direction, suppose that it is not the case that Γ is both complete and consistent. If Γ is not consistent, then by Proposition 2.5.2 it is not satisfiable, which means that it has no truth interpretation. Next suppose that Γ is consistent but is not complete. Then there is some B such that neither B nor $\neg B$ is in Γ. Since Γ is a deductive theory, this means that neither $\Gamma \vdash B$ not $\Gamma \vdash \neg B$. Thus by Lemma 2.6.5, both $\Gamma \cup \{\neg B\}$ and $\Gamma \cup \{B\}$ are consistent. Thus there are truth interpretations I and J such that $\Gamma \cup \{\neg B\}$ satisfies I and $\Gamma \cup \{B\}$ satisfies J. Then $I \neq J$ but Γ satisfies both i and J. So Γ does not have a unique truth interpretation.

$\square$

Recall that the language with two propositional variables A, B has sixteen nonequivalent sentences. If Γ is a consistent complete theory, then for each sentence B, exactly one of $B, \neg B$ is in Γ. It follows that Γ has exactly eight nonequivalent sentences. Each such theory Γ is determined by which of $A, \neg A$ is in Γ and which of $B, \neg B$ is in Γ, and thus corresponds to a single row of the truth table. So there are four different consistent complete theories. $\mathrm{DC}(\{A\})$ has only four nonequivalent sentences $\{A, A \vee B, A \vee \neg B, A \vee \neg A\}$. $\mathrm{DC}(\{A \vee B\})$ has only two sentences, $\{A \vee B, A \vee \neg A\}$.

We conclude this section with several examples of infinite theories.

Example 2.7.7. Let $\mathcal{L} = \{A_0, A_1, \dots\}$.

(1) The set $\Gamma_0 = \{A_0, A_0 \wedge A_1, A_0 \wedge A_1 \wedge A_2, \dots\}$ is complete but not independent.

- It is complete since $\Gamma_0 \vdash A_n$ for all n, which determines the unique truth interpretation I where $I(A_n) = 1$ for all n.
- It is not independent since, for each n, $(A_0 \wedge A_1 \cdots \wedge A_{n+1}) \to (A_0 \wedge \cdots \wedge A_n)$.

(2) The set $\Gamma_1 = \{A_0, A_0 \to A_1, A_1 \to A_2, \dots\}$ is complete and independent. To show that Γ_1 is independent, it suffice to show that, for each single sentence $A_n \to A_{n+1}$, it is not the case that $\Gamma_1 \setminus \{A_n \to A_{n+1}\} \vdash (A_n \to A_{n+1})$. This is witnessed by the interpretation I where $I(A_j) = 1$ if $j \leq n$ and $I(A_j) = 0$ if $j > n$.

(3) The set $\Gamma_2 = \{A_0 \vee A_1, A_2 \vee A_3, A_4 \vee A_5, \dots\}$ is independent but not complete. It is not complete since there are many different interpretations satisfied by Γ_2. In particular, one interpretation could make A_n true if and only if n is odd, and another could make A_n true if and only if n is even.

Note that the last two examples must not be finitely axiomatizable since they are independent.

Exercises for Section 2.7

Exercise 2.7.1. Show that Γ is a deductive theory if and only if $\mathrm{DC}(\Gamma) = \Gamma$.

Exercise 2.7.2. Show that $\mathrm{DC}(\{A \vee B, B \vee C\})$ is not complete in the language with three variables A, B, C.

Exercise 2.7.3. How many nonequivalent sentences are there in a complete consistent theory for the language of three propositional variables A, B, C?

Exercise 2.7.4. How many nonequivalent sentences are there in $\mathrm{DC}(\{A\})$ in the language of three propositional variables A, B, C? What about $A \wedge B$, $A \vee B$?

Exercise 2.7.5. How many different complete consistent theories are there in the language of three propositional variables A, B, C?

Exercise 2.7.6. Show that the first example in Example 2.7.7 is not finitely axiomatizable.

Exercise 2.7.7. Use the Compactness Theorem to show that if Γ is any theory such that $\mathrm{DC}(\Gamma)$ is finitely axiomatizable, then there is a finite subset Δ of Γ such that $\mathrm{DC}(\Delta) = \mathrm{DC}(\Gamma)$.

Exercise 2.7.8. Prove Proposition 2.7.6: Γ is complete and consistent if and only if there is a unique interpretation I satisfied by Γ.

Exercise 2.7.9. Show that part 2 of Example 2.7.7 is complete.

Exercise 2.7.10. Show that part 3 of Example 2.7.7 is independent.

2.8.   Logical vs. Topological Compactness

In this section, we show how the set of truth interpretations may viewed as a topological space, in fact the classical Cantor space $\{0, 1\}^{\mathbb{N}}$. We see that the Compactness Theorem for propositional logic may be seen as showing the compactness of this space.

Fix the set of propositional variables $\mathrm{PVar} = \{A_0, A_1, \dots\}$ and let $\mathcal{I} = \{I : \mathrm{PVar} \to \{0, 1\}\}$ be the set of interpretations of the variables.

Definition 2.8.1.

(1) For any theory Γ, let $\mathrm{Mod}(\Gamma) \subseteq \mathcal{I}$ be the set of interpretations (or *models*) $I : \mathrm{PVar} \to \{0,1\}$ such that I satisfies Γ, that is, $I(P) = 1$ for all $P \in \Gamma$.
(2) The topology on $\mathcal{I}$ has for its basic open sets $\{\mathrm{Mod}(\Gamma) : \Gamma \text{ is finite}\}$. This means that a set $U \subseteq \mathcal{I}$ is open if it is the union of basic open sets, that is, for any $I \in U$, there is some finite set Γ of sentences such that $I \in \mathrm{Mod}(\Gamma)$ and $\mathrm{Mod}(\Gamma) \subseteq U$.

The notation $\mathrm{Mod}(\Gamma)$ is shorthand for the set of *models* of Γ; the term "model" is used here as synonymous with "interpretation".

Note here that the sets Γ may contain arbitrary sentences, whereas we are restricting the interpretations I to the propositional variables. If $\Gamma = \{P\}$ for some sentence P, then we just write $\mathrm{Mod}(P)$ for $\mathrm{Mod}(\{P\})$. It is also important here that the family of basic open sets is countable, since there are only countably many sentences and thus only countably many finite sets of sentences. It follows that every open set is a countable union of basic open sets. Then, as usual, every union of open sets is open.

Lemma 2.8.2.

(1) *For any finite theory Γ, there is a single sentence P such that* $\mathrm{Mod}(\Gamma) = \mathrm{Mod}(P)$.
(2) *For any theory Γ, $\mathcal{I} \setminus \mathrm{Mod}(\Gamma)$ is also a basic open set.*
(3) *For any theories Γ and Δ, $\mathrm{Mod}(\Gamma) \cap \mathrm{Mod}(\Delta) = \mathrm{Mod}(\Gamma \cup \Delta)$.*

The second item implies that every basic open set is also a closed set; such sets are said to be *clopen*. The third item is needed to ensure that the intersection of two open sets is open. The proofs are left as exercises.

Now a closed set in a topological space is simply the complement of an open set. Thus the union of two closed sets is closed and the intersection of any family of closed sets is closed.

Proposition 2.8.3. *A subset K of $\mathcal{I}$ is closed if and only if $K = \mathrm{Mod}(\Gamma)$ for some set Γ of sentences.*

Proof. First, let Γ be an arbitrary theory. We will show that the complement $\mathcal{I} \setminus \mathrm{Mod}(\Gamma)$ is an open set. Suppose that $I \notin \mathrm{Mod}(\Gamma)$. Then there is some sentence $P \in \Gamma$ such that $I(P) = 0$ and therefore $I(\neg P) = 1$. Then I belongs to the basic open set $\mathrm{Mod}(\neg P)$.

Next, suppose that K is a closed set in $\mathcal{I}$ and let $U = \mathcal{I} \setminus K$. Then by Lemma 2.8.2, there is a countable set $\Delta = \{P_0, P_1, \dots\}$ of sentences such that

$$U = \bigcup_n \mathrm{Mod}(P_n).$$

It follows that

$$K = \bigcap_n \mathrm{Mod}(\neg P_n) = \mathrm{Mod}(\{\neg P_0, \neg P_1, \dots\}).$$
$\square$

Now we can show that logical compactness is equivalent to topological compactness. Let us recall the definition of compactness for a topological space.

Definition 2.8.4. A topological space X is compact if, for any sequence $K_0 \supseteq K_1 \supseteq K_2 \cdots$ of closed nonempty sets in X, $\bigcap_n K_n \neq \emptyset$. This is equivalent to saying that for any sequence $U_0 \subseteq U_1 \subseteq U_2 \subseteq \cdots$ of open sets, if $X = \bigcup_n U_n$, then $X = \bigcup_{m<n} U_m$ for some n.

Proposition 2.8.5. *The Compactness Theorem for propositional logic implies that the space $\mathcal{I}$ is compact.*

Proof. First, assume Theorem 2.6.4 and suppose that $K_0 \supseteq K_1 \supseteq K_2 \cdots$ is a sequence of closed nonempty sets in $\mathcal{I}$. Then, for each n, $K_n = \mathrm{Mod}(\Gamma_n)$ for some set Γ_n of sentences by Proposition 2.8.3. Since $K_{n+1} \subseteq K_n$ for all n, we may choose Γ_n so that $\Gamma_0 \subseteq \Gamma_1 \subseteq \dots$. (That is, we can replace each Γ_n with $\Gamma'_n = \bigcup_{m \leq n} \Gamma_m$ since $K_n = \mathrm{Mod}(\Gamma'_n)$ and $\Gamma_n \subseteq \Gamma'_n$ for each

n). For each n, Γ_n is consistent since $\mathrm{Mod}(\Gamma_n) \neq \emptyset$. It follows from Theorem 2.6.4 that $\cup_n \Gamma_n$ is consistent, so that it has an interpretation I. So $I \in K_n$ for each n and therefore $\cap_n K_n \neq \emptyset$. This demonstrates that $\mathcal{I}$ is compact.

For the other direction, assume that the space $\mathcal{I}$ is compact and let the set $\Gamma = \{P_0, P_1, \dots\}$ be given such that, for each n, the set $\Gamma_n = \{P_0, \dots, P_{n-1}\}$ is consistent. Observe that $\mathrm{Mod}(\Gamma) = \cap_n \mathrm{Mod}(\Gamma_n)$. Then $\mathrm{Mod}(\Gamma_{n+1}) \subseteq \mathrm{Mod}(\Gamma_n)$ for each n and each $\mathrm{Mod}(\Gamma_n) \neq \emptyset$, since Γ_n is consistent. It follows from the compactness of $\mathcal{I}$ that the set

$$\mathrm{Mod}(\Gamma) = \bigcap_n \mathrm{Mod}(\Gamma_n) \neq \emptyset,$$

and therefore Γ is consistent. This demonstrates the Compactness Theorem for propositional logic. □

Exercises for Section 2.8

Exercise 2.8.1. Show that for any finite set Γ of sentences, there is a single sentence P such that $\mathrm{Mod}(\Gamma) = \mathrm{Mod}(P)$.

Exercise 2.8.2. Show that for any finite set Γ of sentences, $\mathcal{I} \setminus \mathrm{Mod}(\Gamma)$ is also a basic open set.

Exercise 2.8.3. Show that for any sets Γ and Δ of sentences, $\mathrm{Mod}(\Gamma) \cap \mathrm{Mod}(\Delta) = \mathrm{Mod}(\Gamma \cup \Delta)$.

Exercise 2.8.4. Show that if $K_n = \mathrm{Mod}(\Gamma_n)$ and $K_{n+1} \subseteq K_n$ for each n, then $K_n = \mathrm{Mod}(\cup_{m \leq n} \Gamma_m)$ for each n.

Chapter 3

Predicate Logic

Propositional logic treats a basic part of the language of mathematics, building more complicated sentences from simple with connectives. However, it is inadequate as it stands to express the richness of mathematics. Consider for example the properties of the natural numbers $\mathbb{N} = \{0, 1, 2, \dots\}$ equipped with a linear ordering $<$ and binary operations of addition and multiplication. To say that 0 is the least element, we need to write $(\forall x)\neg(x < 0)$. To say that there is no greatest element, we need to write $(\forall x)(\exists y)x < y$. To test whether an arbitrary structure $(A, <)$ satisfies these properties, we need to know how to interpret $<$ in the set A.

Predicate logic, also called first-order logic, is an enrichment of propositional logic to include predicates, individuals and quantifiers, and is widely accepted as the standard language of mathematics. Axiomatic predicate logic was established by Gottlob Frege [12] in the late 19th century, who formalized the notion of quantified variable which we will present here.

3.1. The Language of Predicate Logic

The symbols of the language of the predicate logic are as follows:

(1) logical connectives, $\neg$, $\vee$, $\wedge$, $\rightarrow$;
(2) the equality symbol $=$;

(3) relation symbols P_i for i in some set I;
(4) function symbols F_j for j in some set J;
(5) constant symbols c_k for k in some set K;
(6) individual variables v_ℓ for each natural number ℓ;
(7) quantifier symbols $\exists$ (the *existential* quantifier) and $\forall$ (the *universal* quantifier);
(8) punctuation symbols (,).

A relation symbol is intended to represent a relation. Thus each relation symbol P is n-ary for some $n \in \mathbb{N}$, which means that we write $P(v_1, \ldots, v_n)$; we say that P is an n-ary or n-place relation symbol. Similarly, a function symbol also is n-ary for some n. A constant symbol may be viewed as a 0-ary function symbol. A first-order language is specified by the choice of relation, function, and constant symbols, so there are many possible first-order languages.

We make a few remarks on the quantifiers:

(a) $(\exists x)\phi$ is read "there exists an x such that ϕ holds".
(b) $(\forall x)\phi$ is read "for all x, ϕ holds".
(c) $(\forall x)\,\theta$ may be thought of as an abbreviation for $(\neg(\exists x)(\neg\theta))$.

Definition 3.1.1. A countable first-order language is obtained by specifying a countable set of relation symbols, function symbols, and constants.

Example 3.1.2. The *language of arithmetic* is specified by $\{S, +, \times, 0, 1\}$. Here S is a unary function symbol, $+$ and $\times$ are 2-place function symbols, and 0, 1 are constants. We often include the 2-place relation symbol $<$. Equality is a special 2-place relation symbol that we include in every language.

One can also work with uncountable first-order languages, but aside from a few examples in Chapter 3, we will primarily work with countable first-order languages. An example of a first-order language is the language of arithmetic.

We now describe how first-order sentences are built up from a given language $\mathcal{L}$. Constants and variables are intended to represent elements of a structure, such as $\mathbb{N}$ with the natural interpretations of 0 and 1. Terms are built up from constants and variables by application of the functions. Here is an inductive definition of the notion of a term.

Definition 3.1.3. The collection of *terms* of a language $\mathcal{L}$ is the smallest set S of strings of symbols such that

(1) each variable and constant is a term;
(2) if $t_1, \ldots, t_n$ are terms and F is an n-place function symbol, then $F(t_1, \ldots, t_n)$ is a term.

The *rank* of a term is defined recursively so that

(1) each variable and constant has rank 0;
(2) if $t_1, \ldots, t_n$ are terms of ranks $r_1, \ldots, r_n$ and F is an n-place function symbol, then the term $F(t_1, \ldots, t_n)$ has rank $1 + \max\{r_1, \ldots, r_n\}$.

A *constant term* is a term with no variables.

Example 3.1.4. The sentences of propositional logic may be viewed as terms in the language $\{\neg, \vee, \wedge, \rightarrow\}$ with one unary and three binary function symbols. Recall that in this language we can show by induction on rank that the number of occurrences of propositional variables in a sentence equals one plus the number of occurrences of binary connectives (see Exercise 1.1.3). Similarly one can show, by induction on the rank of terms, that in the language of arithmetic, the number of occurrences of variables and constants in a term equals one plus the number of occurrences of binary function symbols.

Example 3.1.5. In the language of successor $\{S, 0\}$, the terms will be exactly $\{S^n 0 : n \in \mathbb{N}\} \cup \{S^n v : v \text{ a variable} \wedge n \in \mathbb{N}\}$. This is easy to see by induction on rank. The term $S^n 0$ is intended to represent the natural number n.

Example 3.1.6. In the language $\{S, +, \times, D, 0\}$, where $D(x, y)$ is intended to represent the quotient x/y, there are terms such as $D(S0, SSS0)$, which may be thought of as representing the rational number $\frac{1}{3}$.

Definition 3.1.7. Let $\mathcal{L}$ be a first-order language. The collection of $\mathcal{L}$-formulas is defined by recursion. First, the set of *atomic formulas* consists of formulas of one of the following forms:

(1) $P(t_1, \ldots, t_n)$ where P is an n-place relation symbol and $t_1, \ldots, t_n$ are terms;
(2) $t_1 = t_2$ where t_1 and t_2 are terms.

The set of $\mathcal{L}$-formulas is the smallest set of strings of symbols from $\mathcal{L}$ containing the atomic formulas and closed under the following rules:

(3) If ϕ and θ are $\mathcal{L}$-formulas, then $(\phi \vee \theta)$ is an $\mathcal{L}$-formula. Similarly, $(\phi \wedge \theta)$ and $(\phi \to \theta)$ are $\mathcal{L}$-formulas.
(4) If ϕ is an $\mathcal{L}$-formula, then $(\neg\phi)$ is an $\mathcal{L}$-formula.
(5) If ϕ is an $\mathcal{L}$-formula, then $(\exists v)\phi$ and $(\forall v)\phi$ are $\mathcal{L}$-formulas.

If the language $\mathcal{L}$ is understood from the context, then we speak simply of formulas as opposed to $\mathcal{L}$-formulas.

Example 3.1.8. In the pure language of equality, the following sentence asserts that there are at least two elements:

$$(\exists x)(\exists y)\neg(x = y).$$

Example 3.1.9. In the language of arithmetic, $x + 0 = x$ and $x + x = y$ are atomic formulas. Here are some formulas with quantifiers:

- The formula $(\exists y)(x + y = 0)$ says that x has an additive inverse.
- The formula $(\forall y)(y \times x = y)$ says that x is the multiplicative identity.

- The sentence $(\forall x)(\forall y)(x + y = y + x)$ expresses that addition is commutative.
- The sentence $(\forall x)(\exists y)(x + y = 0)$ says that every element has an additive inverse.
- The sentence $(\forall y)(\exists x)(x + x = y)$ says that every element is divisible by two.

Example 3.1.10. In the language $\{\in\}$ of set theory, the formula $(\forall x)(x \in y \to x \in z)$ says that y is a subset of z.

Example 3.1.11. In the language $\{R\}$ with one binary relation symbol, the sentence $(\forall x)R(x, x)$ says that R is *reflexive*. The sentence $(\forall x)(\forall y)(R(x, y) \to R(y, x))$ says that R is *symmetric*.

In the next section, we shall how formulas may be interpreted in given structures, such as the natural numbers or the real numbers.

An occurrence of a variable v in a formula ϕ becomes *bound* when it is placed in the scope of a quantifier, i.e., $(\exists v)$ or $(\forall v)$ is placed in front of ϕ, and otherwise v is *free*. The formal definition of bound and free variables is given by recursion.

Definition 3.1.12. An occurrence of a variable v is *free* in a formula ϕ if

(1) ϕ is atomic;
(2) ϕ is $(\psi \vee \theta)$ and the occurrence of v is free in whichever one of ψ and θ it appears;
(3) ϕ is $(\neg\psi)$ and the occurrence of v is free in ψ;
(4) ϕ is $(\exists y)\psi$ and the occurrence of v is free in ψ and y is not v.

We say that x is free in the formula ϕ if there is *some* free occurrence of x in ϕ.

Example 3.1.13.

(1) In the atomic formula $x + 5 = 12$, the variable x is free.
(2) In the formula $(\exists x)(x \times x = 2)$, the variable x is bound.

(3) In the formula $x < 7 \lor (\exists x)(2x = 3)$, the first occurrence of x is free but the second occurrence is bound. This is bad form but not technically forbidden.

Definition 3.1.14. A $\mathcal{L}$-formula with no free variables is an $\mathcal{L}$-*sentence*.

In propositional logic, there is no distinction between formulas and sentences.

Definition 3.1.15.

(1) A first-order theory in a language $\mathcal{L}$ is a set Γ of $\mathcal{L}$-sentences.
(2) Γ is said to be *complete* if, for any $\mathcal{L}$-sentence ϕ, either $\phi \in \Gamma$ or $\neg\phi \in \Gamma$.

Exercises for Section 3.1

Exercise 3.1.1. Express the following in the language of equality.

(1) There are exactly two elements.
(2) There are at most three elements.
(3) There are at least three elements.

Exercise 3.1.2. Write a sentence which says that the binary relation R is transitive.

Exercise 3.1.3. Write a formula in the language of set theory which says that $z = x \cup y$ (the union of x and y).

Exercise 3.1.4. Write a sentence in the language of arithmetic which expresses that multiplication is associative.

Exercise 3.1.5. Write a sentence in the language of arithmetic which expresses that every cubic polynomial $x^3 + a_2 x^2 + a_1 x + a_0$, with leading coefficient 1, has a zero.

Exercise 3.1.6. Write a sentence in the language $\{<, 0\}$ which expresses that 0 is the least element.

3.2. Structures

In propositional logic, we use truth tables and interpretations to consider the truth values of complex statements in terms of their simplest components. In predicate logic, to consider the truth values of complex statements that involve quantified variables, we need to introduce *structures*, also called *models*, with universes from which we can select the possible values for the variables.

Consider the formulas from Example 3.1.13. Whether $x+5 = 12$ is true depends on how the variable x is interpreted. If $x = 6$, then this formula is false, whereas if $x = 7$, then it is true. The sentence $(\exists x)(x \times x = 2)$ is true in the set of real numbers, but is not true in the set of rational numbers. The sentence $(\exists x)(2x = 3)$ is true in the rational numbers but is false in the integers.

Definition 3.2.1. Suppose that $\mathcal{L}$ is a first-order language with

(i) relation symbols $P_1, P_2, \ldots$,
(ii) function symbols $F_1, F_2, \ldots$, and
(iii) constant symbols $c_1, c_2, \ldots$.

Then an *$\mathcal{L}$-structure* $\mathcal{A}$ consists of

(a) a nonempty set A (called the *domain* or *universe* of $\mathcal{A}$),
(b) a relation $P_i^{\mathcal{A}}$ on A corresponding to each relation symbol P_i,
(c) a function $F_i^{\mathcal{A}}$ on A corresponding to each function symbol F_i, and
(d) a element $c_i^{\mathcal{A}} \in A$ corresponding to each constant symbol c_i.

Each relation $P_i^{\mathcal{A}}$ requires the same number of places as P_i, so that $P_i^{\mathcal{A}}$ is a subset of A^r for some fixed r (called the *arity* of P_i). In addition, each function $F_i^{\mathcal{A}}$ requires the same number of places as F_i, so that $F_i^{\mathcal{A}} : A^r \to A$ for some fixed r (called the *arity* of F_i).

The relation $P_i^{\mathcal{A}}$ provides an interpretation of the relation symbol P_i, so that $P_i(a_1, \ldots, a_r)$ is *true* in $\mathcal{A}$ if and only if $(a_1, \ldots, a_r) \in P_i^{\mathcal{A}}$. Next, we will see how to extend this notion to arbitrary formulas.

Definition 3.2.2. Given a $\mathcal{L}$-structure $\mathcal{A}$, an *assignment* α into $\mathcal{A}$ is a function α from the variables of $\mathcal{L}$ into the universe A of $\mathcal{A}$.

Just as a truth interpretation for propositional logic may be extended from the variables to the family of all sentences, an assignment here may be extended from the variables to the family of all terms, as follows.

Definition 3.2.3. An assignment α into $\mathcal{A}$ defined on the variables of a language $\mathcal{L}$ may be uniquely extended to the terms of the language by defining, for each function symbol F_i of arity n and any terms $t_1, \ldots, t_n$:

$$\alpha(F_i(t_1, \ldots, t_n)) = F_i^{\mathcal{A}}(\alpha(t_1), \ldots, \alpha(t_n)).$$

Example 3.2.4. Let $\mathcal{L} = \{e, F\}$, where F is a binary function symbol. Let $A = \{a, b, c, d\}$, let $e^{\mathcal{A}} = a$, and let $F^{\mathcal{A}}$ be given by the following table:

$F^{\mathcal{A}}$	a	b	c	d
a	a	b	c	d
b	b	c	d	a
c	c	d	a	b
d	d	a	b	c

Let x, y be variables and let $t = F(x, F(y, e))$. Define the assignment $\alpha : \{x, y, z\} \to A$ by $\alpha(x) = b$, $\alpha(y) = c$, and $\alpha(z) = d$. Then, in the structure $\mathcal{A} = (A, a, F^{\mathcal{A}})$,

$$\alpha(F(y, e)) = F^{\mathcal{A}}(\alpha(y), e^{\mathcal{A}}) = F^{\mathcal{A}}(c, a) = c$$

and
$$\alpha(t) = F^{\mathcal{A}}(\alpha(x), \alpha(F(y, e)) = F^{\mathcal{A}}(b, c) = d.$$

For any assignment α and any variable x and for any element b of the universe, let $\alpha_{b/x}$ be the assignment defined by

$$\alpha_{b/x}(z) = \begin{cases} b & \text{if } z = x, \\ \alpha(z) & \text{otherwise.} \end{cases}$$

Next, we define the notion of truth in a structure.

Definition 3.2.5 (Tarski). We define the relation that a structure $\mathcal{A}$ *satisfies* a formula ϕ via an assignment α into $\mathcal{A}$, denoted by $\mathcal{A} \models \phi[\alpha]$. The definition proceeds by recursion on the complexity of the formula ϕ.

For atomic formulas, we have:

(1) $\mathcal{A} \models (t = s)[\alpha]$ if and only if $\alpha(t) = \alpha(s)$;
(2) $\mathcal{A} \models P_i(t_1, \ldots, t_n)[\alpha]$ if and only if $P_i^{\mathcal{A}}(\alpha(t_1), \ldots, \alpha(t_n))$.

For formulas built up by the logical connectives we have:
(3) $\mathcal{A} \models (\phi \vee \theta)[\alpha]$ if and only if $\mathcal{A} \models \phi[\alpha]$ or $\mathcal{A} \models \theta[\alpha]$;
(4) $\mathcal{A} \models (\phi \wedge \theta)[\alpha]$ if and only if $\mathcal{A} \models \phi[\alpha]$ and $\mathcal{A} \models \theta[\alpha]$;
(5) $\mathcal{A} \models (\phi \rightarrow \theta)[\alpha]$ if and only if $\mathcal{A} \not\models \phi[\alpha]$ or $\mathcal{A} \models \theta[\alpha]$;
(6) $\mathcal{A} \models (\neg\phi)[\alpha]$ if and only if $\mathcal{A} \not\models \phi[\alpha]$.

For formulas built up with quantifiers:

(7) $\mathcal{A} \models ((\exists v)\phi)[\alpha]$ if and only if there is an a in A such that $\mathcal{A} \models \phi[\alpha_{a/v}]$;
(8) $\mathcal{A} \models ((\forall v)\phi)[\alpha]$ if and only if, for every a in A, $\mathcal{A} \models \phi[\alpha_{a/v}]$.

If $\mathcal{A} \models \phi[\alpha]$ for every assignment α, we will simply write $\mathcal{A} \models \phi$. In this case we say that $\mathcal{A}$ is a *model* of ϕ. The *theory* of $\mathcal{A}$ is the set of sentences ϕ such that $\mathcal{A} \models \phi$.

Example 3.2.6. In Example 3.2.4 above, we see that $\mathcal{A} \models$ $(F(x, F(y, e)) = z)[\alpha]$. Now consider the sentences $P : (\exists x)$ $F(x, x) = x$, $Q : (\forall x)F(e, x) = x$, and $R : (\exists y)(\forall x)F(y, x) = x$. To see that $\mathcal{A} \models P$, let α be *any* assignment, that is, let $\alpha(x) \in A$, and observe that $\mathcal{A} \models (F(x, x) = x)[\alpha_{e/x}]$, since $F(e, e) = e$. To see that $\mathcal{A} \models Q$, again let α be any assignment and observe that, for any $d \in A$, $\mathcal{A} \models (F(e, x) = x)[\alpha_{d/x}]$ since $F^A(e, d) = d$ for all $d \in A$. Finally, $\mathcal{A} \models R$ since $\mathcal{A} \models (\forall x)F(e, x) = x$.

Example 3.2.7. In the language $\{S, 0, +\}$, there are terms which are intended to represent constant multiples such as $3v = v + v + v$. We are assuming here that the operation $+$ is associative. Then we can have finite sums of such terms, $c_0 + c_1 v_1 + \cdots + c_n v_n$, for some natural numbers $c_0, \ldots, c_n$. It is not hard to see that, for any term t, there is a term $c_0 + c_1 v_1 + \cdots + c_n v_n$ such that for any assignment α, $\mathbb{N} \models$ $t[\alpha] = c_0 + c_1 \alpha(v_1) + \cdots + c_n \alpha(v_n)$.

Example 3.2.8. For any polynomial $c_0 + c_1 x + c_2 x^2 + \cdots +$ $c_n x^n$ with natural number coefficients, there is a term t in the language $\{S, 0, +, \times\}$ such that for any assignment α into $\mathbb{N}$, $\mathbb{N} \models t[\alpha] = c_0 + c_1 \alpha(x) + c_2 \alpha(x)^2 + \cdots + c_n \alpha(x)^n$.

Proposition 3.2.9. *Let $\mathcal{A}$ be an $\mathcal{L}$-structure and let α and β be any assignments into $\mathcal{A}$.*

(1) *If α and β agree on all the variables occurring in a term t, then $\alpha(t) = \beta(t)$.*
(2) *If α and β agree on all the variables free in a $\mathcal{L}$-formula ϕ, then $\mathcal{A} \models \phi[\alpha]$ if and only if $\mathcal{A} \models \phi[\beta]$.*

Proof. This is a straightforward induction on rank, first on the rank of terms, and then on the rank of formulas. □

Corollary 3.2.10. *If ϕ is a $\mathcal{L}$-sentence, then for all assignments α and β, we have $\mathcal{A} \models \phi[\alpha]$ if and only if $\mathcal{A} \models \phi[\beta]$.*

For $\mathcal{L}$-sentences, we drop the assignment which indicates the interpretation of the variables, and we say simply $\mathcal{A}$ *models* ϕ.

Definition 3.2.11. Let ϕ be an $\mathcal{L}$-formula.

(i) ϕ is *logically valid* if $\mathcal{A} \models \phi[\alpha]$ for every $\mathcal{L}$-structure $\mathcal{A}$ and every assignment α into $\mathcal{A}$.
(ii) ϕ is *satisfiable* if there is some $\mathcal{L}$-structure $\mathcal{A}$ and some assignment α into $\mathcal{A}$ such that $\mathcal{A} \models \phi[\alpha]$.

Definition 3.2.12. Let Γ be any set of formulas in a language $\mathcal{L}$. Let ϕ be a formula in $\mathcal{L}$.

- A $\mathcal{L}$-structure $\mathcal{A}$ is a *model* of Γ if and only if $\mathcal{A} \models Q[\alpha]$ for all $Q \in \Gamma$ and all assignments α;
- Γ is *satisfiable* if there is some $\mathcal{L}$-structure $\mathcal{A}$ and some assignment α such that $\mathcal{A} \models Q[\alpha]$ for all $Q \in \Gamma$.
- Γ implies ϕ, written $\Gamma \models \phi$, if, for any $\mathcal{L}$-structure $\mathcal{A}$ and any assignment α such that $\mathcal{A} \models Q[\alpha]$ for all $Q \in \Gamma$, $\mathcal{A} \models \phi[\alpha]$; if $\Gamma = \{\theta\}$ for some formula θ, then we just say that θ implies ϕ and write $\theta \models \phi$.

In Example 3.1.9, the sentence $(\forall x)(\exists y)(x + y = 0)$, which says that every element has an additive inverse, is true in the integers but false in the natural numbers. The sentence $(\forall y)(\exists x)(x + x = y)$, which says that every element is divisible by two, is true in the rational numbers but false in the integers.

Example 3.2.13. In the language of pure equality, the sentence $(\exists x)(\exists y)(x \neq y)$ is satisfied by a structure $\mathcal{A}$ if and only if $\mathcal{A}$ has at least two distinct elements. More generally, there is a sentence ψ_n such that a structure $\mathcal{A}$ satisfies ψ_n if and only if $\mathcal{A}$ has at least n distinct elements. It is clear that $\psi_{n+1} \models \psi_n$. Let $\Gamma_\infty = \{\psi_n : n \in \mathbb{N}\}$. Then $\mathcal{A}$ is a model of Γ_∞ if and only if $\mathcal{A}$ is infinite.

Example 3.2.14. The language of arithmetic $\{S, +, \times, 0\}$ has natural structures $\mathbb{N}$ (the natural numbers), $\mathbb{Z}$ (the integers), $\mathbb{Q}$ (the rationals), and $\mathbb{R}$ (the reals), with the usual addition and multiplication. Here the successor function S is defined by $S(x) = x + 1$. Sometimes the ordering relation $<$ is included. Consider the two sentences $\phi : (\forall y)(\exists x)(y = Sx)$ and $\psi : (\forall x)(\exists y)(x < y)$. The first sentence is true in $\mathbb{Z}$, $\mathbb{Q}$ and $\mathbb{R}$ but is false in $\mathbb{N}$ since 0 does not have a predecessor x such that $Sx = 0$. The second sentence is true in $\mathbb{N}$, $\mathbb{Z}$, $\mathbb{Q}$ and $\mathbb{R}$.

Exercises for Section 3.2

Exercise 3.2.1. Determine whether the following sentences hold in $\mathbb{N}$, $\mathbb{Z}$, $\mathbb{Q}$ and $\mathbb{R}$:

(a) $(\forall y)(\exists x)(x < y)$;
(b) $(\forall x)(\forall y)(x < y \rightarrow (\exists z)(x < z \wedge z < y)$;
(c) $(\forall x)(\forall y)(x \leq y \leftrightarrow (\exists z)(y = x + z))$.

Exercise 3.2.2. Determine whether the following sentences hold in $\mathbb{N}$, $\mathbb{Z}$, $\mathbb{Q}$, $\mathbb{R}$, and $\mathbb{C}$:

(a) $(\forall y)(\exists x)(y = x + x)$;
(b) $(\forall y)(y > 0 \rightarrow (\exists z)(y = z \times z))$;
(c) $(\forall u_1)(\forall u_2)(\forall u_3)(\exists x)(u_1 x^2 + u_2 x + u_3 = 0)$.

Exercise 3.2.3. Show that, for any term t in the language $\{S, 0, +\}$, there are natural numbers $c_0, \ldots, c_n$ such that, for any assignment α into $\mathbb{N}$, $\mathbb{N} \models t[\alpha] = c_0 + c_1 \alpha(v_1) + \cdots + c_n \alpha(v_n)$.

Exercise 3.2.4. Show that in the language $\{S, +, \times, D, 0\}$, where $D(x, y)$ is intended to represent the quotient x/y, every element of $\mathbb{Q}^+$ (the positive rational numbers) will have a name. That is, for any rational $q > 0$, there is a variable-free term t such that $\mathbb{Q}^+ \models t^{\mathbb{Q}^+} = q$.

Exercise 3.2.5. Show that for any polynomial $c_0 + c_1 x + c_2 x^2 + \cdots + c_n x^n$ with natural number coefficients, there is a term t in the language $\{S, 0, +, \times\}$ such that for any assignment α into $\mathbb{N}$, $\mathbb{N} \models t[\alpha] = c_0 + c_1 \alpha(x) + c_2 \alpha(x)^2 + \cdots + c_n \alpha(x)^n$.

Exercise 3.2.6. For every $\mathcal{L}$-formula ϕ, for all assignments α, β, if α and β agree on all the variables free in ϕ, then $\mathcal{A} \models \phi[\alpha]$ if and only if $\mathcal{A} \models \phi[\beta]$.

3.3. The Deductive Calculus

The predicate calculus is a system of axioms and rules which permit us to derive the true statements of predicate logic without the use of interpretations. The basic relation in the predicate calculus is the relation *proves* between a set Γ of $\mathcal{L}$ formulas and an $\mathcal{L}$-formula ϕ, which formalizes the concept that Γ proves ϕ. This relation is denoted $\Gamma \vdash \phi$. As a first step in defining this relation, we give a list of additional rules of deduction, which extend the list we gave for propositional logic. The logical principles of inference presented here can be traced back to Frege [12].

Some of our rules of the predicate calculus require that we exercise some care in how we substitute variables into certain formulas. For any term s, any variable x and any term t, the term $s[t/x]$ is obtained from s by replacing each free occurrence of x with t. For example, if $s = 2x + y$ and $t = x + y$, then $s[t/x] = 2(x + y) + y$. Similarly, for any formula ϕ, any variable x and any term t, the formula $\phi[t/x]$ is obtained from ϕ by replacing each free occurrence of x with t.

Let us say that $\phi[t/x]$ is a *legal substitution* of t for x in ϕ if no free occurrence of x in ϕ occurs in the scope of a quantifier of any variable appearing in t. For instance, if ϕ has the form $(\forall y)\phi(x, y)$, where x is free, then we cannot legally substitute y in for x, since then y would be bound by the universal quantifier.

Example 3.3.1. If ϕ is the formula $(\exists y)(y \neq x)$, then $\phi[y/x]$ would be the formula $(\exists y)(y \neq y)$, which we expect never to be true. This is not a legal substitution.

(12) (Equality Rule) For any term t, the formula $t = t$ may be derived from Γ in one step.

(13) (Term Substitution) For any terms $t_1, t_2, \ldots, t_n$, $s_1, s_2, \ldots, s_n$, and any function symbol F, if each of the sentences $t_1 = s_1, t_2 = s_2, \ldots, t_n = s_n$ have been derived from Γ, then we may derive $F(t_1, t_2, \ldots, t_n) = F(s_1, s_2, \ldots, s_n)$ from Γ in one additional step.

(14) (Atomic Formula Substitution) For any terms $t_1, t_2, \ldots, t_n$, $s_1, s_2, \ldots, s_n$ and any atomic formula ϕ, if each of the sentences $t_1 = s_1, t_2 = s_2, \ldots, t_n = s_n$, and $\phi(t_1, t_2, \ldots, t_n)$, have been derived from Γ, then we may derive $\phi(s_1, s_2, \ldots, s_n)$ from Γ in one additional step.

(15) ($\forall$-Application) For any term t, if $\phi[t/x]$ is a legal substitution and $(\forall x)\phi$ has been derived from Γ, then we may derive $\phi[t/x]$ from Γ in one additional step.

(16) ($\exists$-Application) If $(\exists x)\phi(x)$ has been derived from Γ and y is a new variable that does not appear free in any formula in Γ or any formula previously derived in our proof, then we may derive $\phi(y/x)$ from Γ in one additional step.

To show that $\Gamma \cup \{(\exists x)\phi(x) \vdash \theta\}$, it suffices to show $\Gamma \cup \{\phi(y/x) \vdash \theta\}$, where y is a new variable that does not appear free in any formula in Γ nor in θ. Then we may derive θ from $\Gamma \cup \{(\exists x)\phi(x)\}$ in one additional step.

(17) ($\forall$-Introduction) Suppose that y does not appear free in any formula in Γ, in any temporary assumption, nor in $(\forall x)\phi$. If $\phi[y/x]$ has been derived from Γ, then we may derive $(\forall x)\phi$ from Γ in one additional step.

(18) ($\exists$-Introduction) If $\phi[t/x]$ is a legal substitution and $\phi[t/x]$ has been derived from Γ, then we may derive $(\exists x)\phi$ from Γ in one additional step.

We remark on three of the latter four rules. First, the reason for the restriction on substitution in ∀-Application is that we need to ensure that t does not contain any free variable that would be become bound when we substitute t for x in ϕ. For example, consider the formula $(\forall x)(\exists y)(x < y)$ in the language of arithmetic. Let ϕ be the formula $(\exists y)(x < y)$, in which x is free but y is bound. Observe that if we substitute the term y for x in ϕ, the resulting formula is $(\exists y)(y < y)$. Thus, from $(\forall x)(\exists y)(x < y)$ we can derive, for instance, $(\exists y)(x < y)$ or $(\exists y)(c < y)$, but we cannot derive $(\exists y)(y < y)$.

Second, the idea behind ∃-Application is this: Suppose in the course of a proof we have derived $(\exists x)\phi(x)$. Informally, we would like to use the fact that ϕ holds of some x to draw some further conclusion θ, but to do so, we need to refer to this object. So we pick an unused variable, say a, and use this as a temporary name to stand for the object satisfying ϕ. Thus, we can write down $\phi(a)$ in a subproof to derive θ. Having derived θ in this way, we can discharge the assumption $\phi(a)$ and conclude θ in our main proof.

Third, in ∀-Introduction, if we think of the variable y as an arbitrary object, then when we show that y satisfies ϕ, we can conclude that ϕ holds of *every* object. However, if y is free in a premise in Γ or a temporary assumption, it is not arbitrary. For example, suppose we begin with the statement $(\exists x)(\forall z)(x + z = z)$ in the language of arithmetic and suppose we derive $(\forall z)(y + z = z)$ by ∃-Application (where y is a temporary name). We are *not* allowed to apply ∀-Introduction here, for otherwise we could conclude $(\forall x)(\forall z)(x + z = z)$, an undesirable conclusion.

Definition 3.3.2. The relation $\Gamma \vdash \phi$ is the smallest subset of pairs (Γ, ϕ) such that Γ is a set of formulas and ϕ is a formula, that contains every pair (Γ, ϕ) such that $\phi \in \Gamma$ or ϕ is $t = t$ for some term t, and which is closed under the eighteen rules of deduction.

As in Propositional Calculus, to demonstrate that $\Gamma \vdash \phi$, we construct a proof.

Example 3.3.3. $\vdash (\exists x)(x = x)$. We have $\vdash x = x$ by the Equality Rule. Then $\vdash (\exists x)(x = x)$ follows by $\exists$-Introduction.

The next example shows that it is possible to remove the universal quantifier from the language, since it can be defined in terms of the existential quantifier.

Example 3.3.4. $\vdash (\forall x)\phi \leftrightarrow \neg(\exists x)\neg\phi$. First assume $(\forall x)\phi(x)$. Now suppose by way of contradiction that $(\exists x)\neg\phi(x)$. Let $\neg\phi(c)$ for some c. Then $\phi(c)$ by $\forall$-Application. This contradiction shows that $\neg(\exists x)\neg\phi(x)$. Thus $\vdash (\forall x)\phi \rightarrow \neg(\exists x)\neg\phi$ by $\rightarrow$-Introduction.

Next assume $\neg(\exists x)\neg\phi$. Let x be arbitrary and suppose by way of contradiction that $\neg\phi(x)$. Then $(\exists x)\neg\phi(x)$ by $\exists$-Introduction. This contradiction shows that $\neg\neg\phi(x)$, by $\neg$-Introduction. Then by $\neg$-Application, we get $\phi(x)$. Since x was arbitrary, it follows by $\forall$-Introduction that $(\forall x)\phi(x)$.

Example 3.3.5. $(\exists x)(\forall y)\theta(x, y) \vdash (\forall y)(\exists x)\theta(x, y)$. Here is a proof which uses all four quantifier rules. As for the deductions in the propositional logic, we will demonstrate this in three ways: First, using the definition of $\vdash$; second, writing out the deduction in sentence form; third, by a formal proof.

Using the definition: Let $\Gamma = \{(\exists x)(\forall y)\theta(x, y)\}$. First, let a be a temporary name such that $(\forall y)\theta(a, y)$ holds. Then by $\forall$-Introduction we have $(\forall y)\theta(a, y) \vdash \theta(a, y)$. Then we get $(\forall y)\theta(a, y) \vdash (\exists x)\theta(x, y)$ by $\exists$-Introduction. Since a does not appear free in Γ nor in $(\exists x)\theta(x, y)$, it follows by $\exists$-Application that $\Gamma \vdash (\exists x)\theta(x, y)$. Since y does not appear free in Γ and there are now no temporary assumptions, it follows by $\forall$-Application that $\Gamma \vdash (\forall y)(\exists x)\theta(x, y)$.

In sentence form: Assume $(\exists x)(\forall y)\theta(x, y)$. Let y be arbitrary. Choose a so that $(\forall y)\theta(a, y)$. Then in particular, $\theta(a, y)$. Hence $(\exists x)\theta(x, y)$. Since y was arbitrary, it follows that $(\forall y)(\exists x)\theta(x, y)$.

A formal proof:

1	$(\exists x)(\forall y)\theta(x, y)$	Given
2	$(\forall y)\theta(a, y)$	Assumption
3	$\theta(a, y)$	$\forall$-Application 2
4	$(\exists x)\,\theta(x, y)$	$\exists$-Introduction 3
5	$(\exists x)\theta(x, y)$	$\exists$-Application 2–4
6	$(\forall y)(\exists x)\,\theta(x, y)$	$\forall$-Introduction 4.

The next proposition exhibits several proofs using the new axiom and rules of predicate logic.

Proposition 3.3.6.

(1) $(\forall x)(\forall y)(x = y \to y = x)$.

(2) $(\forall x)(\forall y)(\forall z)((x = y \,\wedge\, y = z) \to x = z)$.

(3) $(\forall x)\theta(x) \to (\exists x)\theta(x)$.

(4) (i) $(\exists x)(\phi(x) \vee \psi(x)) \vdash (\exists x)\phi(x) \vee (\exists x)\psi(x)$;
   (ii) $(\exists x)\phi(x) \vee (\exists x)\psi(x) \vdash (\exists x)(\phi(x) \vee \psi(x))$.

(5) (i) $(\forall x)(\phi(x) \wedge \psi(x)) \vdash (\forall x)\phi(x) \wedge (\forall x)\psi(x)$;
   (ii) $(\forall x)\phi(x) \wedge (\forall x)\psi(x) \vdash (\forall x)(\phi(x) \wedge \psi(x))$.

(6) $(\exists x)(\phi(x) \wedge \psi(x)) \vdash (\exists x)\phi(x) \wedge (\exists x)\psi(x)$.

(7) $(\forall x)\phi(x) \vee (\forall x)\psi(x) \vdash (\forall x)(\phi(x) \vee \psi(x))$.

(8) $(\forall x)(\phi(x) \to \psi(f(x))) \to ((\exists x)\phi(x) \to (\exists x)\psi(x))$.

Proof.

(1) $(\forall x)(\forall y)(x = y \rightarrow y = x)$.

1	$x = y$	Assumption
2	$x = x$	Equality Rule
3	$y = x$	Atomic Formula
4		Substitution 1,2
5	$x = y \rightarrow y = x$	$\rightarrow$-Introduction 1–3
6	$(\forall y)(x = y \rightarrow y = x)$	$\forall$-Introduction 4
7	$(\forall x)(\forall y)(x = y \rightarrow y = x)$	$\forall$-Introduction 5.

(6) $(\exists x)(\phi(x) \wedge \psi(x)) \vdash (\exists x)\phi(x) \wedge (\exists x)\psi(x)$

1	$(\exists x)(\phi(x) \wedge \psi(x))$	Given
2	$\phi(a) \wedge \psi(a)$	Assumption
3	$\phi(a)$	$\wedge$-Application 2
4	$(\exists x)\phi(x)$	$\exists$-Introduction 3
5	$\psi(a)$	$\wedge$-Application 2
6	$(\exists x)\psi(x)$	$\exists$-Introduction 5
7	$(\exists x)\phi(x) \wedge (\exists x)\psi(x)$	$\wedge$-Introduction 4, 6
8	$(\exists x)\phi(x) \wedge (\exists x)\psi(x)$	$\exists$-Application 2–7.

In sentence form: Suppose that $(\exists x)(\phi(x) \wedge \psi(x))$. Choose a so that $\phi(a) \wedge \psi(a)$. Then $\phi(a)$, so that $(\exists x)\phi(x)$. Also, $\psi(a)$, so that $(\exists x)\psi(x)$. Putting the two together, we have $(\exists x)\phi(x) \wedge (\exists x)\psi(x)$. Since a was a temporary name, we can thus conclude $(\exists x)\phi(x) \wedge (\exists x)\psi(x)$.

(7) $(\forall x)\,\phi(x) \vee (\forall x)\,\psi(x) \vdash (\forall x)(\phi(x) \vee \psi(x))$

1	$(\forall x)\,\phi(x) \vee (\forall x)\,\psi(x)$	Given
2	$(\forall x)\,\phi(x)$	Assumption
3	$\phi(x)$	$\forall$-Application 2
4	$\phi(x) \vee \psi(x)$	$\vee$-Introduction 3
5	$(\forall x)(\phi(x) \vee \psi(x))$	$\forall$-Introduction 4
6	$(\forall x)\,\psi(x)$	Assumption
7	$\psi(x)$	$\forall$-Application 6
8	$\phi(x) \vee \psi(x)$	$\vee$-Introduction 7
9	$(\forall x)(\phi(x) \vee \psi(x))$	$\forall$-Introduction 8
10	$(\forall x)(\phi(x) \vee \psi(x))$	$\vee$-Application 1–9.

In sentence form: We are given $(\forall x)\,\phi(x) \vee (\forall x)\,\psi(x)$. Let x be arbitrary. There are two cases. First, suppose that $(\forall x)\phi(x)$. Then $\phi(x)$. Second, suppose that $(\forall x)\psi(x)$. Then $\psi(x)$. In either case, we have $\phi(x) \vee \psi(x)$. Since x was arbitrary, this implies $(\forall x)(\phi(x) \vee \psi(x))$.

(8) $(\forall x)(\phi(x) \to \psi(f(x))) \vdash (\exists x)\phi(x) \to (\exists x)\psi(x)$.

1	$(\forall x)(\phi(x) \to \psi(f(x)))$	Given
2	$(\exists x)\phi(x)$	Assumption
3	$\phi(a)$	Assumption
4	$\phi(a) \to \psi(f(a))$	$\forall$-Application 1
5	$\psi(f(a))$	$\to$-Application 3, 4
6	$(\exists x)\,\psi(x)$	$\exists$-Introduction 5
7	$(\exists x)\,\psi(x)$	$\exists$-Application 3–6
8	$(\exists x)\phi(x) \to (\exists x)\psi(x)$	$\to$-Introduction 2–6 $\square$

We can now carry out some easy proofs from naive set theory in our deductive system. Here we use the language $\{\in, \cup, \cap, \setminus\}$, and use the following set Γ^{set} of definitions/axioms:

(i) $x \in A \cup B \leftrightarrow (x \in A \vee x \in B)$;
(ii) $x \in A \cap B \leftrightarrow (x \in A \wedge x \in B)$;
(iii) $x \in A \setminus B \leftrightarrow (x \in A \wedge \neg x \in B)$;
(iv) $A = B \leftrightarrow (\forall x)(x \in A \leftrightarrow x \in B)$;
(v) $A \subseteq B \leftrightarrow (\forall x)(x \in A \rightarrow x \in B)$.

Example 3.3.7. We prove from Γ^{set} that $A \cup B = B \cup A$. By item (iv), we need to show that $(\forall x)(x \in A \cup B \leftrightarrow x \in B \cup A)$. Let x be arbitrary and suppose that $x \in A \cup B$. Then $x \in A \vee x \in B$, by item (i). It follows from part (5) of Proposition 1.4.5 that $x \in B \vee x \in A$, so that $x \in B \cup A$. Hence $x \in A \cup B \rightarrow x \in B \cup A$. A similar argument shows that $x \in B \cup A \rightarrow x \in A \cup B$. Thus $x \in A \cup B \leftrightarrow x \in B \cup A$. Since x was arbitrary, $(\forall x)(x \in A \cup B \leftrightarrow x \in B \cup A)$. Thus by item (iv), $A \cup B = B \cup A$.

The following notion will be useful.

Definition 3.3.8. A formula ϕ is in *prenex form* if it consists of a sequence of quantifiers followed by a quantifier-free formula.

Proposition 3.3.9. *Any first-order formula ϕ is logically equivalent to a formula in prenex form.*

Proof. The proof is by induction on rank. Clearly adding a quantifier at the front of a formula keeps it in prenex form. For the logical connectives, it suffices to show the induction step for negation and disjunction. These are left as exercises. □

Example 3.3.10. The sentence $(\exists x)P(x) \wedge (\exists x)Q(x)$ is logically equivalent to $(\exists x)(\exists y)(P(x) \wedge Q(y))$.

Exercises for Section 3.3

Exercise 3.3.1. Give an informal proof that the sentence $(\exists x)P(x) \wedge (\exists x)Q(x)$ is logically equivalent to $(\exists x)(\exists y)(P(x) \wedge Q(y))$.

Exercise 3.3.2. Here are some cases of the proof that any formula ϕ is logically equivalent to a prenex formula.

(a) Show that for any prenex formula θ, $\neg\theta$ is logically equivalent to a prenex formula by induction on the number of quantifiers at the front of θ.

(b) Show that for any prenex formulas θ and ψ, $(\exists x)\theta \wedge (\exists y)\psi$ is logically equivalent to a prenex formula.

Exercise 3.3.3. Let ϕ be the formula $(\exists y)(x+y = z)$. For each term t below, write the formula $\phi[t/x]$ and determine whether t is a legal substitution for x in ϕ.

(1) $x + z$;
(2) $x + y$;
(3) z.

Exercise 3.3.4. Give a proof in sentence form that $(\forall x)(\forall y)(x = y \rightarrow y = x)$.

Exercise 3.3.5. Give a formal proof that $(\forall x)(\forall y)(\forall z)((x = y \wedge y = z) \rightarrow x = z)$.

Exercise 3.3.6. Give a formal proof that $(\forall x)\theta(x) \rightarrow (\exists x)\theta(x)$.

Exercise 3.3.7. Give a formal proof that $(\exists x)(\phi(x) \vee \psi(x)) \vdash (\exists x)\phi(x) \vee (\exists x)\psi(x)$.

Exercise 3.3.8. Give a formal proof that $(\exists x)\phi(x) \vee (\exists x)\psi(x) \vdash (\exists x)(\phi(x) \vee \psi(x))$.

Exercise 3.3.9. Give a formal proof that $(\forall x)(\phi(x) \wedge \psi(x)) \vdash (\forall x)\phi(x) \wedge (\forall x)\psi(x)$.

Exercise 3.3.10. Give a formal proof that $(\forall x)\phi(x) \wedge (\forall x)\psi(x) \vdash (\forall x)(\phi(x) \wedge \psi(x))$.

Exercise 3.3.11. Give a proof in sentence form that $(\forall x)(\phi(x) \to \psi(f(x))) \vdash (\exists x)\phi(x) \to (\exists x)\psi(x)$.

Exercise 3.3.12. Prove that, for any sets A, B, and C, $A \cup (B \cap C) = (A \cup B) \cap (A \cup C)$.

Exercise 3.3.13. Prove that, for any sets A, B, and C, $A \cup (B \cup C) = (A \cup B) \cup C$.

Exercise 3.3.14. Prove that, for any sets A, B, and C, $A \setminus (B \cup C) = (A \setminus B) \cup (A \setminus C)$.

3.4. Soundness Theorem for Predicate Logic

Our next goal is to prove the soundness of our system of proofs. The following lemma is needed in the proof.

Lemma 3.4.1. *Let $\mathcal{L}$ be a first-order language and let $\mathcal{B}$ be an $\mathcal{L}$-structure. For every variable x, every term t, and every assignment α in B, if $\phi[t/x]$ is a legal substitution, then*

(1) *for every term r, if $b = \alpha(t)$, then $\alpha(r[t/x]) = \alpha_{b/x}(r)$;*
(2) *for every $\mathcal{L}$-formula ϕ,*

$$\mathcal{B} \models \phi[t/x][\alpha] \quad \textit{if and only if} \quad \mathcal{B} \models \phi[\alpha_{b/x}],$$

where $b = \alpha(t)$.

Proof. The proof is by routine induction on the rank of terms and then on the rank of formulas. $\qquad\square$

Theorem 3.4.2 (Soundness Theorem of Predicate Logic). *Let Γ be a set of formulas and ϕ a formula. If $\Gamma \vdash \phi$, then $\Gamma \models \phi$.*

Proof. As in the proof of the soundness theorem for propositional logic, the proof is again by induction on the length of the deduction of ϕ. We need to show that if there is a proof of ϕ from Γ, then for any structure $\mathcal{A}$ and any assignment α into $\mathcal{A}$, if $\mathcal{A} \models \gamma[\alpha]$ for all $\gamma \in \Gamma$, then $\mathcal{A} \models \phi[\alpha]$. The arguments for the rules from propositional logic carry over here, so we just need to verify the result holds for the new rules.

Suppose the result holds for all formulas obtained in proofs of length strictly less than n lines.

- (Equality Rule) Suppose the last line of a proof of length n with premises Γ is $t = t$ for some term t and let α be any assignment. Then since $\alpha(t) = \alpha(t)$, we have $\mathcal{A} \models (t = t)[\alpha]$.
- (Term Substitution) Suppose the last line of a proof of length n with premises Γ is $F(s_1, \ldots, s_n) = F(t_1, \ldots, t_n)$, obtained by term substitution. Then we must have established $s_1 = t_1, \ldots, s_n = F(t_n)$ earlier in the proof. By the inductive hypothesis, we have $\Gamma \models s_1 = t_1, \ldots, \Gamma \models s_n = t_n[\alpha]$. Suppose that $\mathcal{A} \models \gamma$ for every $\gamma \in \Gamma$. Then $\alpha(s_i) = \alpha(t_i)$ for $i = 1, \ldots, n$. So

$$\alpha(F(s_1, \ldots, s_n)) = F^{\mathcal{A}}(\alpha(s_1), \ldots, \alpha(s_n))$$

$$= F^{\mathcal{A}}(\alpha(t_1), \ldots, \alpha(t_n))$$

$$= \alpha(F(t_1), \ldots, (t_n)).$$

Hence $\mathcal{A} \models (F(s_1, \ldots, s_n) = F(t_1, \ldots, t_n))[\alpha]$. Since $\mathcal{A}$ and α were arbitrary, we can conclude that $\Gamma \models \phi[t/x]$.

- (Atomic Formula Substitution) The argument is similar to the previous one and is left to the reader.
- ($\forall$-Application) *For any term t, if $\phi[t/x]$ is a legal substitution and $(\forall x)\phi$ has been derived from Γ, then we may derive $\phi[t/x]$ from Γ in one additional step.*

  Suppose that the last line of a proof of length n with premises Γ is $\phi[t/x]$, obtained by $\forall$-Application. Thus, we must have derived $\forall x \phi(x)$ earlier in the proof. Let $\mathcal{A} \models \gamma[\alpha]$ for every $\gamma \in \Gamma$. Then by the inductive hypothesis, we have

$\mathcal{A} \models \forall x \phi(x)[\alpha]$, which implies that $\mathcal{A} \models \phi[\alpha_{a/x}]$ for every $a \in A$. If $\alpha(t) = b$, then since $\mathcal{A} \models \phi[\alpha_{b/x}]$, by Lemma 3.4.1 we have $\mathcal{A} \models \phi[t/x][\alpha]$. Since $\mathcal{A}$ and α were arbitrary, we can conclude that $\Gamma \models \phi[t/x]$.

- ($\exists$-Application) *To show that $\Gamma \cup \{(\exists x)\phi(x)\} \vdash \theta$, it suffices to show $\Gamma \cup \{\phi[y/x]\} \vdash \theta$, where y is a new variable that does not appear free in any formula in Γ nor in θ.*

  Suppose that the last line of a proof of length n with premises Γ is given by $\exists$-Application. Then $\Gamma \vdash (\exists x)\phi(x)$ in less than n lines and $\Gamma \cup \{\phi[y/x]\} \vdash \theta$ in less than n lines. Let $\mathcal{A} \models \gamma[\alpha]$ for every $\gamma \in \Gamma$. Then by the inductive hypothesis, we have $\Gamma \models (\exists x)\phi[\alpha]$, which implies that $\mathcal{A} \models \phi[\alpha_{b/x}]$ for some $b \in A$. Let $\beta = \alpha_{b/y}$, so that $\beta(y) = b$. It follows that $\mathcal{A} \models \phi[\beta]$, since $\alpha = \beta$ except on possibly y and y does not appear free in ϕ. Then by Lemma 3.4.1, $\mathcal{A} \models \phi[y/x][\beta]$, and hence $\mathcal{A} \models \theta[\beta]$. It follows that $\mathcal{A} \models \theta[\alpha]$, again since $\alpha = \beta$ except on possibly y. Since $\mathcal{A}$ and α were arbitrary, we can conclude that $\Gamma \cup (\exists x)\phi \models \theta$.

- ($\forall$-Introduction) *Suppose that y does not appear free in any formula in Γ, in any temporary assumption, nor in $(\forall x)\phi$. If $\phi[y/x]$ has been derived from Γ, then we may derive $(\forall x)\phi$ from Γ in one additional step.*

  Suppose that the last line of a proof of length n with premises Γ is $(\forall x)\phi(x)$, obtained by $\forall$-Introduction. Thus, we must have derived $\phi[y/x]$ from Γ earlier in the proof, where y satisfies the necessary conditions described above. Let $\mathcal{A} \models \gamma[\alpha]$ for every $\gamma \in \Gamma$. Since y does not appear free in Γ, then for any $a \in A$, $\mathcal{A} \models \Gamma[\alpha_{a/y}]$. For an arbitrary $a \in A$, let $\beta = \alpha_{a/y}$, so that $\beta(y) = a$. By induction, we have $\Gamma \models \phi[y/x]$, which implies that $\mathcal{A} \models \phi[y/x][\beta]$. Then by Lemma 3.4.1, $\mathcal{A} \models \phi[\beta_{a/x}]$. Since $\alpha_{a/x} = \beta_{a/x}$ except on possibly y, which does not appear free in ϕ, we have $\mathcal{A} \models \phi[\alpha_{a/x}]$. As a was arbitrary, we have shown $\mathcal{A} \models \phi[\alpha_{a/x}]$ for every $a \in A$. Hence $\mathcal{A} \models (\forall x)\phi(x)[\alpha]$. Since $\mathcal{A}$ and α were arbitrary, we can conclude that $\Gamma \models (\forall x)\phi(x)$.

- (∃-Introduction) *If $\phi[t/x]$ is a legal substitution and $\phi[t/x]$ has been derived from* Γ, *then we may derive* $(\exists x)\phi$ *from* Γ *in one additional step.*

  Left as an exercise. □

Exercises for Section 3.4

Exercise 3.4.1. Show closure under the Atomic Formula Substitution rule in the proof of the Soundness Theorem.

Exercise 3.4.2. Show closure under the ∃-Introduction rule in the proof of the Soundness Theorem.

Chapter 4

Models of Predicate Logic

In this chapter, we will prove Gödel's Completeness Theorem for predicate logic by showing how to build a model for a consistent first-order theory.

Gödel first proved this result in 1929 [13] in his doctoral dissertation. We will also discuss several consequences of the Compactness Theorem for first-order logic and consider several relations that hold between various models of a given first-order theory, namely isomorphism and elementary equivalence.

For a more thorough introduction to the subject of model theory, books by Marker [25], Hodges [19], and Chang and Keisler [6].

4.1. The Completeness Theorem for Predicate Logic

Fix a first-order language $\mathcal{L}$. For convenience, we will assume that our $\mathcal{L}$-formulas are built up only using $\neg$, $\vee$, and $\exists$.

We will also make use of the following key facts (the proofs of which we omit):

(1) If A is a tautology in propositional logic, then if we replace each instance of each propositional variable in A with an $\mathcal{L}$-formula, the resulting $\mathcal{L}$-formula is true in all $\mathcal{L}$-structures.

(2) For any $\mathcal{L}$-structure $\mathcal{A}$ and any assignment α into $\mathcal{A}$,

$$\mathcal{A} \models ((\forall x)\phi)[\alpha] \Leftrightarrow \mathcal{A} \models (\neg(\exists x)\neg\phi)[\alpha].$$

We will also use the following analogs of results we proved in Chapter 2, the proofs of which are the same:

Lemma 4.1.1. *Let Γ be an $\mathcal{L}$-theory.*

(1) *For an $\mathcal{L}$-sentence ϕ, $\Gamma \vdash \phi$ if and only if $\Gamma \cup \{\neg\phi\}$ is inconsistent.*
(2) *If Γ is consistent, then for any $\mathcal{L}$-sentence ϕ, either $\Gamma \cup \{\phi\}$ is consistent or $\Gamma \cup \{\neg\phi\}$ is consistent.*

The following result, known as the Constants Theorem, plays an important role in the proof of the completeness theorem.

Theorem 4.1.2 (Constants Theorem). *Let Γ be an $\mathcal{L}$-theory and let c be a constant symbol which does not occur in $\mathcal{L}$. Let $\phi(x)$ be an $\mathcal{L}$-formula. If $\Gamma \vdash \phi(c/x)$, then $\Gamma \vdash (\forall x)\phi(x)$. Furthermore, if Γ is consistent as an $(\mathcal{L} \setminus \{c\})$-theory, then Γ is still consistent as an $(\mathcal{L} \cup \{c\})$-theory.*

Proof. Given a proof of $\phi(c)$ from Γ, let v be a variable not appearing in Γ. If we replace every instance of c with v in the proof of $\phi(c)$, we have a proof of $\phi(v)$ from Γ. Then by $\forall$-Introduction, we have $\Gamma \vdash (\forall x)\phi(x)$. The second part has a similar proof. □

Gödel's Completeness Theorem can be articulated in two ways, which we will prove are equivalent.

Theorem 4.1.3 (Completeness Theorem, Version I). *For any $\mathcal{L}$-theory Γ and any $\mathcal{L}$-sentence ϕ,*

$$\Gamma \models \phi \Rightarrow \Gamma \vdash \phi.$$

Theorem 4.1.4 (Completeness Theorem, Version II). *Every consistent $\mathcal{L}$-theory has a model.*

We claim that the two versions are equivalent.

Proof of claim. First, suppose that every consistent theory has a model, and suppose further that $\Gamma \models \phi$. If Γ is not consistent, then Γ proves every sentence, and hence $\Gamma \vdash \phi$. If, however, Γ is consistent, we have two cases to consider. If $\Gamma \cup \{\neg\phi\}$ is inconsistent, then by Lemma 4.1.1(2), it follows that $\Gamma \vdash \phi$. In the case that $\Gamma \cup \{\neg\phi\}$ is consistent, by the second version of the Completeness Theorem, there is some $\mathcal{L}$-structure $\mathcal{A}$ such that $\mathcal{A} \models \Gamma \cup \{\neg\phi\}$, from which it follows that $\mathcal{A} \models \Gamma$ and $\mathcal{A} \models \neg\phi$. But we have assumed that $\Gamma \models \phi$, and hence $\mathcal{A} \models \phi$, which is impossible. Thus, if Γ is consistent, it follows that $\Gamma \cup \{\neg\phi\}$ is inconsistent.

For the other direction, suppose the first version of the Completeness Theorem holds and let Γ be an arbitrary $\mathcal{L}$-theory. Suppose Γ has no model. Then vacuously, $\Gamma \models \phi$ for any sentence ϕ. It follows from the first version of the Completeness Theorem that $\Gamma \vdash \phi$ for any ϕ, and hence Γ is inconsistent. $\square$

We now turn to the proof of the second version of the Completeness Theorem. As in the proof of the Completeness Theorem for propositional logic, we will use the Compactness Theorem, which comes in several forms (just as it did in with propositional logic).

The following notion will be needed.

Definition 4.1.5. Let $\mathcal{L}'$ be an extension of the language $\mathcal{L}$, that is, $\mathcal{L}'$ contains some additional constant symbols, relation symbols, or function symbols. If $\mathcal{A}'$ is an $\mathcal{L}'$ structure, then the *reduct* $\mathcal{A}$ of $\mathcal{A}'$ to the language $\mathcal{L}$ is the $\mathcal{L}$-structure with the same universe as $\mathcal{A}'$, which has the same interpretations as $\mathcal{A}'$ for the symbols of $\mathcal{L}$.

Theorem 4.1.6. *Let Γ be an $\mathcal{L}$-theory.*

(1) *For an $\mathcal{L}$-sentence ϕ, if $\Gamma \vdash \phi$, there is some finite $\Gamma_0 \subseteq \Gamma$, $\Gamma_0 \vdash \phi$.*

(2) *If every finite $\Gamma_0 \subseteq \Gamma$ is consistent, then Γ is consistent.*

(3) *If* $\Gamma = \bigcup_n \Gamma_n$ *with* $\Gamma_n \subseteq \Gamma_{n+1}$ *for all* n *and each* Γ_n *is consistent, then* Γ *is consistent.*

As in the case of propositional logic, (1) follows by induction on proof length, while (2) follows directly from (1) and (3) follows directly from (2).

Our strategy for proving the Completeness Theorem is as follows. Given Γ, we want to extend it to a maximally consistent collection of $\mathcal{L}$-formulas, like the proof of the Completeness Theorem for propositional logic. The problem that we now encounter (that did not occur in the propositional case) is that it is unclear how to handle sentences of the form $(\exists x)\theta$.

The solution to this problem, due to Henkin [14], is to extend the language $\mathcal{L}$ to a language $\mathcal{L}^+$ by adding new constants $c_0, c_1, c_2, \ldots$, which we will use to witness the truth of existential sentences.

Definition 4.1.7. A set Δ of formulas in a language $\mathcal{L}$ is *Henkin complete* if, for each $\mathcal{L}^+$-formula $\theta(v) \in \Delta$ with exactly one free variable v, if $(\exists v)\theta(v)$ is in Δ, then there is some constant c in the language such that $\theta(c) \in \Delta$.

Hereafter, let us assume that $\mathcal{L}$ is countably infinite (which is not a necessary restriction), so that we will only need to add countably many new constants to our language. Using these constants, we will build a model of Γ, where the universe of our model consists of certain equivalence classes on the set of all $\mathcal{L}^+$-terms with no variables, which will make up the universe of the desired structure. The model will satisfy a collection $\Delta \supseteq \Gamma$ that is maximally consistent and Henkin complete.

Proof of Theorem 4.1.4. Let $\phi_0, \phi_1, \ldots$ be an enumeration of all $\mathcal{L}^+$-sentences. We define a sequence $\Gamma = \Delta_0 \subseteq \Delta_0 \subseteq \Delta_1 \subseteq \cdots$ such that for each $n \in \mathbb{N}$,

$$\Delta_{2n+1} = \begin{cases} \Delta_{2n} \cup \{\phi_n\} & \text{if } \Delta_{2n} \cup \{\phi_n\} \text{ is consistent,} \\ \Delta_{2n} \cup \{\neg\phi_n\} & \text{otherwise} \end{cases}$$

and

$$\Delta_{2n+2} = \begin{cases} \Delta_{2n+1} \cup \{\theta(c_m)\} & \text{if } \phi_n \text{ is of the form } (\exists v)\theta(v) \\ & \text{and is in } \Delta_{2n+1}, \\ \Delta_{2n+1} & \text{otherwise} \end{cases}$$

where c_m is the first constant in our list of new constants that has not appeared in Δ_{2n+1}. Then we define $\Delta = \bigcup_n \Delta_n$.

We now prove a series of claims.

Claim 1. Δ *is complete (that is, for every $\mathcal{L}^+$-sentence ϕ, either $\phi \in \Delta$ or $\neg\phi \in \Delta$).*

Proof of Claim 1. This follows immediately from the construction.

Claim 2. *Each Δ_k is consistent.*

Proof of Claim 2. We prove this claim by induction on k. First, $\Delta_0 = \Gamma$ is consistent by assumption and the Constants Theorem. Now suppose that Δ_k is consistent. If $k = 2n$ for some n, then clearly Δ_{k+1} is consistent, since if $\Delta_{2n} \cup \{\phi_n\}$ is consistent, then we set $\Delta_{k+1} = \Delta_{2n} \cup \{\phi_n\}$, and if not, then by Lemma 4.1.1(3), $\Delta_{2n} \cup \{\neg\phi_n\}$ is consistent, and so we set $\Delta_{k+1} = \Delta_{2n} \cup \{\neg\phi_n\}$.

If $k = 2n+1$ for some n, then if ϕ_n is not of the form $(\exists v)\theta(v)$ or if it is but it is not in Δ_{2n+1}, then $\Delta_{2n+2} = \Delta_{2n+1}$ is consistent by induction. If ϕ_n is of the form $(\exists v)\theta(v)$ and is in Δ_{2n+1}, then let $c = c_m$ be the first constant not appearing in Δ_{2n+1}. Suppose that $\Delta_{k+1} = \Delta_{2n+2} = \Delta_{2n+1} \cup \{\theta(c)\}$ is not consistent. Then by Lemma 4.1.1(2), $\Delta_{2n+1} \vdash \neg\theta(c)$. Then by the Constants Theorem, $\Delta_{2n+1} \vdash (\forall x)\neg\theta(x)$. But since ϕ_n is the formula $(\exists v)\theta(v)$ and is in Δ_{2n+1}, it follows that Δ_{2n+1} is inconsistent, contradicting our inductive hypothesis. Thus $\Delta_{k+1} = \Delta_{2n+2}$ is consistent.

Claim 3. $\Delta = \bigcup_n \Delta_n$ *is consistent.*

Proof of Claim 3. This follows from the part (3) of Theorem 4.1.6.

Claim 4. Δ *is Henkin complete* (*that is, for each $\mathcal{L}^{+}$-formula* $\theta(v)$ *with exactly one free variable and* $(\exists v)\theta(v) \in \Delta$*, we have* $\theta(c) \in \Delta$ *for some constant c*).

Proof of Claim 4. Suppose that $(\exists v)\theta(v) \in \Delta$. Then there is some n such that $(\exists v)\theta(v)$ is the formula ϕ_n. Since $\Delta_{2n} \cup \{\phi_n\} \subseteq \Delta$ is consistent, $(\exists v)\theta(v) \in \Delta_{2n+1}$. Then by construction, $\theta(c) \in \Delta_{2n+2}$ for some constant c.

Our final task is to build a model $\mathcal{A}$ such that $\mathcal{A} \models \Delta$, from which it will follow that $\mathcal{A} \models \Gamma$ (since $\Gamma \subseteq \Delta$). We define an equivalence relation on the Herbrand universe of $\mathcal{L}^{+}$ (i.e., the set of constant $\mathcal{L}^{+}$-terms, or equivalently, the $\mathcal{L}^{+}$-terms that contain no variables). For constant terms s and t, we define

$$s \sim t \Leftrightarrow s = t \in \Delta.$$

Claim 5. $\sim$ *is an equivalence relation.*

Proof of Claim 5. This is left as an exercise.

For a constant term s, let $[s]$ denote the equivalence class of s. Then we define an $\mathcal{L}^{+}$-structure as follows:

(i) $A = \{[t] : t$ is a constant term of $\mathcal{L}^{+}\}$;
(ii) for each function symbol f of the language $\mathcal{L}$, we define

$$f^{\mathcal{A}}([t_1], \ldots, [t_n]) = [f(t_1, \ldots, t_n)],$$

where n is the arity of f;
(iii) for each predicate symbol P of the language $\mathcal{L}$, we define

$$P^{\mathcal{A}}([t_1], \ldots, [t_n]) \text{ if and only if } P(t_1, \ldots, t_n) \in \Delta,$$

where n is the arity of P;
(iv) for each constant symbol c of the language $\mathcal{L}^{+}$, we define

$$c^{\mathcal{A}} = [c].$$

Claim 6. $\mathcal{A} = (A, f^{\mathcal{A}}, \ldots, P^{\mathcal{A}}, \ldots, c^{\mathcal{A}}, \ldots)$ *is well-defined.*

Proof of Claim 6. We have to show in particular that the interpretation of function symbols and predicate symbols in $\mathcal{A}$ is well-defined. Suppose that $s_1 = t_1, \ldots, s_n = t_n \in \Delta$ and

$$f^{\mathcal{A}}([t_1], \ldots, [t_n]) = [f(t_1, \ldots, t_n)]. \tag{4.1}$$

By our first assumption, it follows that $\Delta \vdash s_i = t_i$ for $i = 1, \ldots, n$. Then by term substitution, $\Delta \vdash f(s_1, \ldots, s_n) = f(t_1, \ldots, t_n)$, and so $f(s_1, \ldots, s_n) = f(t_1, \ldots, t_n) \in \Delta$. It follows that

$$[f(s_1, \ldots, s_n)] = [f(t_1, \ldots, t_n)]. \tag{4.2}$$

Combining (4.1) and (4.2) yields

$$f^{\mathcal{A}}([t_1], \ldots, [t_n]) = [f(t_1, \ldots, t_n)]$$
$$= [f(s_1, \ldots, s_n)] = f^{\mathcal{A}}([s_1], \ldots, [s_n]).$$

A similar argument shows that the interpretation of predicate symbols is well-defined.

Claim 7. *Let α be an assignment into $\mathcal{A}$. Then $\alpha(t) = [t]$ for every constant term t.*

Proof of Claim 7. We verify this by induction on the rank of the term t.

- Suppose t is a constant symbol c. Then $\alpha(c) = c^{\mathcal{A}} = [c]$.
- Suppose that t is the term $f(t_1, \ldots, t_n)$ for function symbol f and constant terms $t_1, \ldots, t_n$, where $\alpha(t_i) = [t_i]$ for $i = 1, \ldots, n$ by induction. Then

$$\alpha(f(t_1, \ldots, t_n)) = f^{\mathcal{A}}(\alpha(t_1), \ldots, \alpha(t_n))$$
$$= f^{\mathcal{A}}([t_1], \ldots, [t_n]) = [f(t_1, \ldots, t_n)]. \tag{4.3}$$

Claim 8: $\mathcal{A} \models \Delta$.

Proof of Claim 8. We verify this by proving that for every assignment α into $\mathcal{A}$ and every $\mathcal{L}^+$-sentence ϕ, $\mathcal{A} \models \phi[\alpha]$ if and only if $\phi \in \Delta$.

- If ϕ is $s = t$ for some terms s, t, then

$$
\begin{aligned}
\mathcal{A} \models (s = t)[\alpha] &\Leftrightarrow \alpha(s) = \alpha(t) \\
&\Leftrightarrow [s] = [t] \\
&\Leftrightarrow s \sim t \\
&\Leftrightarrow s = t \in \Delta.
\end{aligned}
$$

- If ϕ is $P(t_1, \ldots, t_n)$ for some predicate symbol P, then

$$
\begin{aligned}
\mathcal{A} \models P(t_1, \ldots, t_n)[\alpha] &\Leftrightarrow P^{\mathcal{A}}(\alpha(t_1), \ldots, \alpha(t_n)) \\
&\Leftrightarrow P^{\mathcal{A}}([t_1], \ldots, [t_n]) \\
&\Leftrightarrow P(t_1, \ldots, t_n) \in \Delta.
\end{aligned}
$$

- If ϕ is $\neg\psi$ for some $\mathcal{L}^+$-sentence ψ, then

$$
\begin{aligned}
\mathcal{A} \models \neg\psi[\alpha] &\Leftrightarrow \mathcal{A} \not\models \psi \\
&\Leftrightarrow \psi \notin \Delta \\
&\Leftrightarrow \neg\psi \in \Delta.
\end{aligned}
$$

- If ϕ is $\psi \vee \theta$ for some $\mathcal{L}^+$-sentences ψ and θ, then

$$
\begin{aligned}
\mathcal{A} \models (\psi \vee \theta)[\alpha] &\Leftrightarrow \mathcal{A} \models \psi[\alpha] \text{ or } \mathcal{A} \models \theta[\alpha] \\
&\Leftrightarrow \psi \in \Delta \text{ or } \theta \in \Delta \\
&\Leftrightarrow \psi \vee \theta \in \Delta.
\end{aligned}
$$

- If ϕ is $(\exists v)\theta(v)$ for some $\mathcal{L}^+$-formula θ with one free variable v, then

$$
\begin{aligned}
\mathcal{A} \models (\exists v)\theta(v)[\alpha] &\Leftrightarrow \mathcal{A} \models \theta(b) \text{ for some } b \in A \\
&\Leftrightarrow \theta(c) \in \Delta, \text{ where } b = [c] \\
&\Leftrightarrow (\exists v)\theta(v) \in \Delta.
\end{aligned}
$$

Since $\mathcal{A} \models \Delta$, it follows that $\mathcal{A} \models \Gamma$. Note that $\mathcal{A}$ is an $\mathcal{L}^+$-structure while Γ is only an $\mathcal{L}$-theory (as it does not contain any expression involving any of the additional constants). Then let $\mathcal{A}^*$ be the reduct of $\mathcal{A}$ to the language $\mathcal{L}$, that is it has the same universe as $\mathcal{A}$ and the same interpretations of the function symbols and predicate symbols of $\mathcal{L}$. Then clearly $\mathcal{A}^* \models \Gamma$, and the proof is complete. □

Exercises for Section 4.1

Exercise 4.1.1. Show that for any $\mathcal{L}$-structure $\mathcal{A}$ and any assignment α into $\mathcal{A}$,

$$\mathcal{A} \models ((\forall x)\phi)[\alpha] \Leftrightarrow \mathcal{A} \models (\neg(\exists x)\neg\phi)[\alpha].$$

Exercise 4.1.2. Show that the relation $\sim$ defined in the proof of Theorem 4.1.4 is an equivalence relation.

Exercise 4.1.3. Finish the proof of Claim 6 by showing that the interpretation of predicate symbols is well-defined.

Exercise 4.1.4. Show that adding new symbols to a language $\mathcal{L}$ does not affect the consistency of a set Γ of sentences in the original language or the provability of sentences in $\mathcal{L}$ from Γ.

Exercise 4.1.5. Show by induction on proof length that for an $\mathcal{L}$-sentence ϕ, if $\Gamma \vdash \phi$, there is some finite $\Gamma_0 \subseteq \Gamma$, $\Gamma_0 \vdash \phi$. *Hint:* See the proof of Theorem 1.4.8.

4.2. Compactness

The same consequences we derived from the Soundness and Completeness Theorem for propositional logic apply now to predicate logic with basically the same proofs.

Theorem 4.2.1. *For any first-order theory Γ, Γ is satisfiable if and only if Γ is consistent.*

Theorem 4.2.2. *If Σ is a consistent theory, then Σ is included in some complete, consistent theory.*

We also have an additional version of the Compactness Theorem, which is the most common formulation of compactness.

Theorem 4.2.3 (Compactness Theorem for Predicate Logic). *An $\mathcal{L}$-theory Γ is satisfiable if and only if every finite subset of Γ is satisfiable.*

Proof.   ($\Rightarrow$) If $\mathcal{A} \models \Gamma$, then it immediately follows that $\mathcal{A} \models \Gamma_0$ for any finite $\Gamma_0 \subseteq \Gamma$.

($\Leftarrow$) Suppose that Γ is not satisfiable. By the Completeness Theorem, Γ is not consistent. Then $\Gamma \vdash \bot$, so that, by the first formulation of the Compactness Theorem, there is some finite $\Gamma_0 \subseteq \Gamma$ such that $\Gamma_0 \vdash \bot$. It follows that Γ_0 is not satisfiable.   □

We now consider some applications of the Compactness Theorem, the first yielding a model of arithmetic with infinite natural numbers and the second yielding a model of the real numbers with infinitesimals.

Example 4.2.4. Let $\mathcal{L} = \{+, \times, <, 0, 1\}$ be the language of arithmetic augmented by $<$, and let Γ be the set of $\mathcal{L}$-sentences true in the standard model of arithmetic. Let us expand $\mathcal{L}$ to $\mathcal{L}'$ by adding a new constant c to our language. We extend Γ to an $\mathcal{L}'$-theory Γ' by adding all sentences of the form

$$\psi_n : c > \underbrace{1 + \cdots + 1}_{n \text{ times}}.$$

We claim that every finite $\Gamma_0' \subseteq \Gamma'$ is satisfiable. Given any finite $\Gamma_0' \subseteq \Gamma'$, Γ_0' consists of at most finitely many sentences from Γ and at most finitely many sentences of the form ψ_i. It follows that

$$\Gamma_0' \subseteq \Gamma \cup \{\psi_{n_1}, \psi_{n_2}, \ldots, \psi_{n_k}\}$$

for some $n_1, n_2, \ldots, n_k \in \mathbb{N}$, where these latter sentences assert that c is larger than each of the values $n_1, n_2, \ldots, n_k$.

Let $n = \max\{n_1, \ldots, n_k\}$ then let $\mathcal{A} = (\mathbb{N}, +, \times, <, 0, 1, n)$, so that $c^{\mathcal{A}} = n$ and hence $\mathcal{A} \models \Gamma_0'$. Then by the Compactness Theorem, there is some $\mathcal{L}'$-structure $\mathcal{B}$ such that $\mathcal{B} \models \Gamma'$. In the universe of $\mathcal{B}$, we have objects that behave exactly like $0, 1, 2, 3, \ldots$ (in a sense we will make precise shortly), but the interpretation of c in $\mathcal{B}$ satisfies $c^{\mathcal{B}} > n$ for every $n \in \mathbb{N}$ and hence behaves like an infinite natural number. We will write the universe of $\mathcal{B}$ as $\mathbb{N}^*$.

Example 4.2.5. Let $\mathcal{L}$ consist of the language of arithmetic together with function symbols for the standard calculus functions $e^x, \sin(x)$, and $\cos(x)$. Let $\mathcal{R}$ be the standard $\mathcal{L}$-structure with universe $\mathbb{R}$ and let $\Gamma_{\mathcal{R}}$ be the set of $\mathcal{L}$-sentences true in the standard model of the real numbers. Let $\mathcal{L}' = \mathcal{L} \cup \{c\} \cup \{c_r : r \in \mathbb{R}\}$ and let $\Gamma' = \Gamma_{\mathcal{R}} \cup \{c > n \times 1 : n \in \mathbb{N}\}$. As in the previous example, every finite $\Gamma_0' \subseteq \Gamma'$ is satisfiable. Hence by the Compactness Theorem, Γ' is satisfiable. Let $\mathcal{A} \models \Gamma'$. The universe of $\mathcal{A}$ contains a copy of $\mathbb{R}$ and also contains an infinite number $c^{\mathcal{A}}$. In addition, $\mathcal{A}$ will contain *infinitesimal* elements, that is, since $c^{\mathcal{A}} > 0$, $c^{\mathcal{A}}$ will have a multiplicative inverse d and $d < 1/n$ for each $n \in \mathbb{N}$. We will write the universe of $\mathcal{A}$ as $\mathbb{R}^*$.

The language of pure equality provides a very interesting illustration of the use of compactness. Let $\mathcal{E}^{\geq n}$ be the sentence stating that there are at least n distinct elements, that is,

$$(\exists x_1)(\exists x_2) \cdots (\exists x_n)(x_1 \neq x_2 \wedge x_1 \neq x_3 \wedge x_2 \neq x_3 \wedge \cdots \wedge x_{n-1} \neq x_n).$$

The conjunct inside the quantifiers can be abbreviated as $\bigwedge_{i \neq j} x_i \neq x_j$. Note that $\mathcal{E}^{\geq n+1}$ implies $\mathcal{E}^{\geq n}$ for each n.

It is clear that a structure $\mathcal{A}$ is infinite if and only if $\mathcal{A} \models \mathcal{E}^{\geq n}$ for all n. Let $\Gamma_\infty = \{\mathcal{E}^{\geq 1}, \mathcal{E}^{\geq 2}, \ldots\}$; Γ_∞ is the *theory of infinity*. Then a structure $\mathcal{A}$ is infinite if and only if $\mathcal{A} \models \mathcal{E}^{\geq n}$ for all n.

Proposition 4.2.6. *There is no first-order sentence γ such that a structure $\mathcal{A}$ is infinite if and only if $\mathcal{A} \models \gamma$.*

Proof. Suppose by way of contradiction that γ is a sentence such that $\mathcal{A}$ is infinite if and only if $\mathcal{A} \models \gamma$. Then $\Gamma_\infty \models \gamma$ and hence, by the Compactness Theorem, there must be a finite subset $\Gamma_n = \{\mathcal{E}^{\geq 1}, \ldots, \mathcal{E}^{\geq n}\}$ of Γ such that $\Gamma_n \models \gamma$. Now consider the structure $\{1, 2, \ldots, n\}$, which satisfies $\mathcal{E}^{\geq n}$ but is not infinite, and hence does not satisfy γ. This contradicts the statement above that $\Gamma_n \models \gamma$. □

Now we consider a question that was not appropriate to consider in the context of propositional logic, namely, what are the sizes of models of a given theory? Our main theorem is a consequence of the proof of the Completeness Theorem. We proved the Completeness Theorem only in the case of a countable language $\mathcal{L}$, and we built a countable model (which was possibly finite). By using a little care (and some set theory), one can modify steps (1) and (2) for an uncountable language to define by transfinite recursion a theory Δ and prove by transfinite induction that Δ has the desired properties. The construction leads to a model whose size is at most the size of the language with which one started. Thus, we have the following theorem from [24].

Theorem 4.2.7 (Downward Löwenheim–Skolem Theorem). *Assume that the first-order language $\mathcal{L}$ is infinite and let Γ be an $\mathcal{L}$-theory with an infinite model. Then Γ has a model of size $\leq |\mathcal{L}|$.*

Proof. Let the structure $\mathcal{A}$ be given with universe A. We will construct a subset B of A of cardinality $\leq |\mathcal{L}|$ and show that for any formula ϕ and any $b_1, \ldots, b_k \in B$, $\mathcal{A} \models \phi(b_1, \ldots, b_k)$ if and only if $\mathcal{B} \models \phi(b_1, \ldots, b_k)$, where $\mathcal{B}$ is the structure with universe B and with the same constants, relations, and functions as $\mathcal{A}$. We will define the universe $B = \bigcup_n B_n$, where B_n is an increasing sequence of subsets of A. Let $B_0 = \emptyset$ and for each n, define B_{n+1} as follows. For each formula $\phi(x, x_1, \ldots, x_k)$ with $k + 1$ free variables, if $\mathcal{A} \models (\exists x)\phi(x, b_1, \ldots, b_k)$, for some $b_1, \ldots, b_n \in B_n$, then we put an element $b \in \mathcal{A}$ into B_{n+1} such that $\mathcal{A} \models \phi(b, b_1, \ldots, b_k)$. Thus $0 < |B_1| \leq |\mathcal{L}|$ and it follows by induction that $|B_{n+1}| \leq |B_n| \cdot |\mathcal{L}| \leq |\mathcal{L}|$ for each n. Therefore $|B| \leq |\mathcal{L}|$.

Now we can prove, by induction on formulas, that for any formula ϕ and any $b_1, \ldots, b_k \in B$, $\mathcal{A} \models \phi(b_1, \ldots, b_k)$ if and only if $\mathcal{B} \models \phi(b_1, \ldots, b_k)$. For atomic formulas, this is true because $\mathcal{B}$ has the same constants, relations, and functions as $\mathcal{A}$. The key induction step is when ϕ is $(\exists x)\psi(x, b_1, \ldots, b_k)$. If ϕ is true in $\mathcal{B}$, then $\mathcal{B} \models \psi(b, b_1, \ldots, b_k)$ for some $b \in \mathcal{B}$ and by induction $\mathcal{A} \models \psi(b, b_1, \ldots, b_k)$, so that $\mathcal{A} \models \phi$. If $\mathcal{A} \models \phi$, let k be large enough so that each $b_1, \ldots, b_k \in B_n$. Then by the construction there exists $b \in B_{n+1}$ such that $\mathcal{A} \models \psi(b, b_1, \ldots, b_k)$. It follows by induction that $\mathcal{B} \models \psi(b, b_1, \ldots, b_k)$, and therefore $\mathcal{B} \models \phi$. $\square$

Corollary 4.2.8. *There is a model $\mathcal{A}$ of nonstandard analysis with cardinality equal to $|\mathbb{R}|$.*

Proof. Let $\mathcal{A}$ be a model of nonstandard analysis given by the compactness argument above. Since there is a constant symbol c_r for each $r \in \mathbb{R}$ and certainly these have distinct interpretations, then $|\mathcal{A}| \geq |\mathbb{R}|$. Now use Theorem 4.2.7 to obtain a subset B of A and a corresponding model $\mathcal{B}$ of nonstandard analysis which has cardinality exactly $|\mathbb{R}|$. $\square$

Theorem 4.2.9 (Upward Löwenheim–Skolem Theorem).
If Γ is an $\mathcal{L}$-theory with an infinite model, then Γ has a model of size κ for every infinite κ with $|\mathcal{L}| \leq \kappa$.

Proof Sketch. First we add κ new constant symbols $\langle d_\alpha : \alpha < \kappa \rangle$ to our language $\mathcal{L}$. Next we expand Γ to Γ' by adding formulas that say $d_\alpha \neq d_\beta$ for the different constants:

$$\Gamma' = \Gamma \cup \{ \neg d_\alpha = d_\beta : \alpha < \beta < \kappa \}.$$

Since Γ has an infinite model, each finite $\Gamma'_0 \subseteq \Gamma'$ has a model. Hence by the Compactness Theorem, Γ' has a model. By the Soundness Theorem, Γ' is consistent. Then use the proof of the

Completeness Theorem to define a model $\mathcal{B}'$ of Γ' the universe of which has size $|B| \leq \kappa$. Since $\mathcal{B}' \models d_\alpha \neq d_\beta$ for $\alpha \neq \beta$, there are at least κ many elements. Thus $|B| = \kappa$ and so Γ' has a model $\mathcal{B}'$ of the desired cardinality. Let $\mathcal{B}$ be the reduct of $\mathcal{B}'$ obtained by removing the new constant symbols from our expanded language. Then $\mathcal{B}$ is a model of the desired size for Γ.   □

The theory of groups from abstract algebra will provide important examples of structures.

Definition 4.2.10. A *group* $\mathcal{A} = (A, +, 0)$ is a set together with an operation $+$ (addition), and an identity element 0, which satisfy the following axioms:

(1) Associative Law: $(\forall x)(\forall y)(\forall z)(x + (y + z) = (x + y) + z)$;
(2) Identity Law: $(\forall x)(x + 0 = x)$;
(3) Inverse Law: $(\forall x)(\exists y)(x + y = 0)$.

The group $\mathcal{A}$ is *commutative* (or *Abelian*) if it also satisfies:

(4) Commutative Law: $(\forall x)(\forall y)(x + y = y + x)$.

Example 4.2.11. The standard structures of the integers $(\mathbb{Z}, +, 0)$, the rationals $(\mathbb{Q}, +, 0)$, and the real numbers $(\mathbb{R}, +, 0)$ are commutative groups. The set $\mathbb{Q}^+$ of *positive* rationals is a commutative group under multiplication with identity 1, but the set of positive integers is not. The set of invertible 2×2 real matrices $\mathcal{M}_2 = (M, *, I)$, with matrix multiplication $*$ and the usual identity matrix I, is a group but is not commutative. Hence the commutative law is independent of the other three laws.

Exercises for Section 4.2

Exercise 4.2.1. Let n be a positive integer and g an element of a group $\mathcal{G} = (G, *, e)$. We say that g has order $n > 0$ if $g^n = e$ but $g^m \neq e$ for any positive $m < n$. Then g has order ∞ if

$g^n \neq e$ for any positive natural number n. Show that there is no formula ϕ of group theory such that $|g| = \infty$ in a group $\mathcal{G}$ if and only if $\mathcal{G} \models \phi(g)$.

Exercise 4.2.2. Use the fact that there are finite commutative groups of order 2^n in which every element has order 2 to show that there must be an infinite commutative group such that every element has order 2.

4.3. Isomorphism and Elementary Equivalence

In this section, we introduce the notions of isomorphism and elementary equivalence, as well as submodels and elementary submodels.

Definition 4.3.1. Given $\mathcal{L}$-structures $\mathcal{A}$ and $\mathcal{B}$, a bijection $H : \mathcal{A} \to \mathcal{B}$ is an isomorphism if it satisfies:

(1) For every constant $c \in \mathcal{L}$, $H(c^{\mathcal{A}}) = c^{\mathcal{B}}$.
(2) For every k-ary predicate symbol $P \in \mathcal{L}$ and every $a_1, \ldots, a_k \in A$,

$$P^{\mathcal{A}}(a_1, \ldots, a_k) \Leftrightarrow P^{\mathcal{B}}(H(a_1), \ldots, H(a_k)).$$

(3) For every k-ary function symbol $F \in \mathcal{L}$ and every $a_1, \ldots, a_k \in A$,

$$H(F^{\mathcal{A}}(a_1, \ldots, a_k)) = F^{\mathcal{B}}(H(a_1), \ldots, H(a_k)).$$

Furthermore, $\mathcal{A}$ and $\mathcal{B}$ are isomorphic, denoted $\mathcal{A} \cong \mathcal{B}$, if there exists an isomorphism between $\mathcal{A}$ and $\mathcal{B}$.

Example 4.3.2. The ordered group $(\mathbb{R}, +, <)$ of real numbers under addition is isomorphic to the ordered group $(\mathbb{R}^{>0}, \times, <)$ of positive real numbers under multiplication under the mapping $H(x) = 2^x$. The key observation here is that $H(x + y) = 2^{x+y} = 2^x \times 2^y = H(x) \times H(y)$.

We compare the relation of isomorphism with the following relation between models.

Definition 4.3.3. $\mathcal{L}$-structures $\mathcal{A}$ and $\mathcal{B}$ are *elementarily equivalent*, denoted $\mathcal{A} \equiv \mathcal{B}$, if for any $\mathcal{L}$-sentence ϕ,

$$\mathcal{A} \models \phi \Leftrightarrow \mathcal{B} \models \phi.$$

How do the relations of $\cong$ and $\equiv$ compare? First, we have the following theorem.

Theorem 4.3.4. *If $\mathcal{A}$ and $\mathcal{B}$ are $\mathcal{L}$-structures satisfying $\mathcal{A} \cong \mathcal{B}$, then $\mathcal{A} \equiv \mathcal{B}$.*

The proof is by induction on the complexity of $\mathcal{L}$-sentences. The converse of this theorem does not hold, as shown by the following example.

Example 4.3.5. $(\mathbb{Q}, \leq)$ and $(\mathbb{R}, \leq)$, both models of the theory of dense linear orders without endpoints, are elementarily equivalent, which follows from the fact that the theory of dense linear orders without endpoints is complete (which we will prove in Chapter 6). Note, however, that these structures are not isomorphic, since they have different cardinalities.

Definition 4.3.6. Let $\mathcal{A}$ and $\mathcal{B}$ be $\mathcal{L}$-structures with corresponding domains $A \subseteq B$.

(1) $\mathcal{A}$ is a *submodel* of $\mathcal{B}$ ($\mathcal{A} \subseteq \mathcal{B}$) if the following are satisfied:
  (a) for each constant $c \in \mathcal{L}$, $c^{\mathcal{A}} = c^{\mathcal{B}}$;
  (b) for each n-ary function symbol $f \in \mathcal{L}$ and each $a_1, \ldots, a_n \in A$,

$$f^{\mathcal{A}}(a_1, \ldots, a_n) = f^{\mathcal{B}}(a_1, \ldots, a_n);$$

  (c) for each n-ary relation symbol $R \in \mathcal{L}$ and each $a_1, \ldots, a_n \in A$,

$$R^{\mathcal{A}}(a_1, \ldots, a_n) \Leftrightarrow R^{\mathcal{B}}(a_1, \ldots, a_n).$$

(2) $\mathcal{A}$ is an *elementary submodel* of $\mathcal{B}$ (written $\mathcal{A} \precsim \mathcal{B}$) if

  (a) $\mathcal{A}$ is a submodel of $\mathcal{B}$;

  (b) for each $\mathcal{L}$-formula $\phi(x_1, \ldots, x_n)$ and each $a_1, \ldots,$
    $a_n \in A$,

$$\mathcal{A} \models \phi(a_1, \ldots, a_n) \Leftrightarrow \mathcal{B} \models \phi(a_1, \ldots, a_n).$$

Here are some examples.

Example 4.3.7. Consider the rings $(\mathbb{Z}, 0, 1, +, \times), (\mathbb{Q}, 0, 1, +, \times)$, and $(\mathbb{R}, 0, 1, +, \times)$.

- $(\mathbb{Z}, 0, 1, +, \times)$ is a submodel of $(\mathbb{Q}, 0, 1, +, \times)$ and $(\mathbb{Q}, 0, 1, +, \times)$ is a submodel of $(\mathbb{R}, 0, 1, +, \times)$.
- $(\mathbb{Z}, 0, 1, +, \times)$ is not an elementary submodel of $(\mathbb{Q}, 0, 1, +, \times)$, since $\mathbb{Q} \models (\exists x)x + x = 1$ which is false in $\mathbb{Z}$.
- Neither $(\mathbb{Z}, 0, 1, +, \times)$ nor $(\mathbb{Q}, 0, 1, +, \times)$ is an elementary submodel of $(\mathbb{R}, 0, 1, +, \times)$ since $\mathbb{R} \models (\exists x)x \times x = 2$, which is false in both $\mathbb{Z}$ and $\mathbb{Q}$.

Example 4.3.8. The following elementary submodel relations hold:

- $(\mathbb{Q}, \leq) \precsim (\mathbb{R}, \leq)$.
- $(\mathbb{N}, 0, 1, +, \times) \precsim (\mathbb{N}^*, 0, 1, +, \times)$.
- $(\mathbb{R}, 0, 1, +, \times) \precsim (\mathbb{R}^*, 0, 1, +, \times)$.

The latter two items in the previous example justify the claims that the natural numbers are contained in models of nonstandard arithmetic and that the real numbers are contained in models of nonstandard analysis.

Here is an example of a subset which is *not* the universe of a submodel.

Example 4.3.9. $\mathbb{Z}_3 = \{0, 1, 2\}$ with addition modulo 3 is not a submodel of $\mathbb{Z}_6 = \{0, 1, 2, 3, 4, 5\}$ with addition modulo 6 because they have different addition functions: In $\mathbb{Z}_3$, $2 + 2 = 1$ whereas in $\mathbb{Z}_6$, $2 + 2 = 4$. However, $\mathbb{Z}_3$ *is* isomorphic to the subgroup of $\mathbb{Z}_6$ consisting of $\{0, 2, 4\}$.

If we take a look at the proof of the Downward and Upward Löwenheim–Skolem Theorems 4.2.7 and 4.2.9, we can see that the following were actually shown:

Theorem 4.3.10 (Löwenheim–Skolem Theorem).

(1) *Assume that $|\mathcal{L}|$ is infinite and let $\mathcal{B}$ be an infinite model of a $\mathcal{L}$-theory Γ. Then $\mathcal{B}$ has a an elementary submodel $\mathcal{A}$ of size $\leq |\mathcal{L}|$.*

(2) *If $\mathcal{A}$ is an infinite model of a $\mathcal{L}$-theory Γ, and κ is a cardinal number such that both $|A|$ and $|\mathcal{L}|$ are $\leq \kappa$, then there is a structure $\mathcal{B}$ of size κ such that $\mathcal{A}$ is an elementary submodel of $\mathcal{B}$.*

Proof.   The downward version is immediate from the proof of Theorem 4.2.7. For the upward version, just add to the language names c_a for each $a \in \mathcal{A}$ and let Γ' be the theory of $\mathcal{A}$ in this language, so that $\Gamma \subseteq \Gamma'$. Use the Upward Löwenheim–Skolem Theorem to obtain a model $\mathcal{B}$ of cardinality κ which satisfies Γ'. Now suppose that $\mathcal{B} \models (\exists x)\phi(x, a_1, \ldots, a_k)$ and let $c_i = c_{a_i}$. Then $\psi = (\exists x)\phi(x, c_1, \ldots, c_k) \in \Gamma'$, so that $\mathcal{A} \models \psi$. Thus the reduct of $\mathcal{B}$ to the language $\mathcal{L}$ also satisfies ψ. The reverse implication always holds.

Just as every subset X of a group G generates a subgroup $\langle X \rangle$ of G, every subset X of an arbitrary structure generates a substructure $\langle X \rangle$.

Example 4.3.11. In $(\mathbb{Z}, +, -)$, $\langle \{20, 30\} \rangle = \langle 10 \rangle = \{10x : x \in \mathbb{Z}\}$.

Example 4.3.12. In the Boolean algebra $\mathcal{B} = (\mathcal{P}(\{1, 2, 3, 4\}), \cap, \cup, -)$, $\langle \{\{2, 3\}, \{4\}\} \rangle$ contains the sets

$$\emptyset, \{1\}, \{4\}, \{1, 4\}, \{2, 3\}, \{1, 2, 3\}, \{2, 3, 4\}, \{1, 2, 3, 4\}.$$

Example 4.3.13. In $(\mathbb{R}, 0, 1, +, -, \times)$, $\langle \sqrt{2} \rangle = \{m + n\sqrt{2} : m, n \in \mathbb{Z}\}$.

Definition 4.3.14. Let $\mathcal{A}$ be a structure with universe A for some language $\mathcal{L}$ and let X be a subset of A. Then $\langle X \rangle$ is the smallest substructure of $\mathcal{A}$ which includes X.

The following proposition is immediate from the definition of submodel.

Proposition 4.3.15. *For any $\mathcal{L}$-structure $\mathcal{A}$ with universe A and any subset B of A, $\langle B \rangle$ is a submodel of $\mathcal{A}$, under the interpretations of $\mathcal{A}$.*

Proposition 4.3.16. *For any structure $\mathcal{A}$ with universe A and any subset B of A, $\langle B \rangle$ is the intersection of the family of submodels of $\mathcal{A}$ which which have B as a subset.*

Proof. Let C be the intersection of the universes of all submodels of $\mathcal{A}$ which include B. By Proposition 4.3.15, there is such a structure with universe $\langle B \rangle$, so that $C \subseteq \langle B \rangle$. For the other direction, let $\mathcal{D}$ be any substructure of $\mathcal{A}$ with universe D which includes B as a subset. We must have $\langle B \rangle \subseteq D$, since D is closed under all functions and hence all terms. It follows that $\langle B \rangle \subseteq C$. $\qquad\qquad\square$

A submodel $\mathcal{A}$ of a structure $\mathcal{B}$ satisfies the same quantifier-free formulas $\phi(a_1, \ldots, a_n)$ as $\mathcal{B}$. If $\mathcal{A}$ is an elementary submodel, then it satisfies the same first order formulas $\phi(a_1, \ldots, a_n)$ as $\mathcal{B}$. Next we consider some intermediate versions of this notion, where $\mathcal{A}$ agrees with $\mathcal{B}$ on a certain class of formulas.

Definition 4.3.17. A first-order formula ϕ is said to be *universal* if there is a quantifier-free formula θ and variables $y_1, \ldots, y_m$ such that

$$\phi = (\forall y_1)(\forall y_2) \cdots (\forall y_m)\theta,$$

and is said to be *logically universal* if it is logically equivalent to a universal formula. Existential and logically existential formulas are similarly defined using existential quantifiers $(\exists y_i)$. A formula ϕ is said to be existential-universal (or $\exists\forall$ for short) if there is a universal formula θ and variables $y_1, \ldots, y_m$ such that $\phi = (\exists y_1)(\exists y_2)\cdots(\exists y_m)\theta$, and is said to be *logically* $\exists\forall$ if it is logically equivalent to a $\exists\forall$ formula. Universal-existential formulas are similarly defined.

Example 4.3.18. The statement that an ordering $<$ has no greatest element can be written

$$(\forall y)(\exists x)(y < x)$$

and is therefore seen to be universal-existential.

Example 4.3.19. The axioms for an equivalence relation in the language $\{E\}$ with one binary relation are universal.

(a) (Reflexive): $(\forall x)xRx$.
(b) (Symmetric): $(\forall x)(\forall y)(xRy \rightarrow yRx)$.
(c) (Transitive): $(\forall x)(\forall y)(\forall z)((xRy \wedge yRz) \rightarrow xRz)$.

The importance of universal sentences is in the following notion of persistence.

Definition 4.3.20. A sentence ϕ is said to be *downward persistent* if whenever $\mathcal{A} \subseteq \mathcal{B}$ and $\mathcal{B} \models \phi$, then $\mathcal{A} \models \phi$. This can also apply to a formula $\phi(x_1, \ldots, x_n)$ where $\mathcal{B} \models \phi(a_1, \ldots, a_n)$ implies that $\mathcal{A} \models \phi(a_1, \ldots, a_n)$ when $a_1, \ldots, a_n \in A$. Similarly, ϕ is *upward persistent* if whenever $\mathcal{A} \subseteq \mathcal{B}$ and $\mathcal{A} \models \phi(a_1, \ldots, a_n)$, then $\mathcal{B} \models \phi(a_1, \ldots, a_n)$.

It is immediate from these definitions that ϕ is upward persistent if and only if $\neg\phi$ is downward persistent.

Proposition 4.3.21. *Any universal formula ϕ is downward persistent.*

Proof. Let $\phi(x_1, \ldots, x_n)$ have the form

$$(\forall y_1)(\forall y_2) \cdots (\forall y_m)\theta(x_1, \ldots, x_n, y_1, \ldots, y_m),$$

where θ is quantifier-free. Suppose that $\mathcal{A} \subseteq \mathcal{B}$ and that $\mathcal{B} \models \phi(a_1, \ldots, a_n)$, where $a_1, \ldots, a_n \in A$. Now let $c_1, \ldots, c_n \in A$ be arbitrary. Then

$$\mathcal{B} \models \theta(a_1, \ldots, a_n, c_1, \ldots, c_m),$$

since $\mathcal{B} \models (\forall y_1)(\forall y_2) \cdots (\forall y_m)\theta(a_1, \ldots, a_n, y_1, \ldots, y_m)$. Since $\mathcal{A} \subseteq \mathcal{B}$ and θ is quantifier-free, it follows that $\mathcal{A} \models \theta(a_1, \ldots, a_n, c_1, \ldots, c_m)$. Since $c_1, \ldots, c_m$ are arbitrary elements of A, we may now conclude that

$$\mathcal{A} \models (\forall y_1)(\forall y_2) \cdots (\forall y_m)\theta(a_1, \ldots, a_n, y_1, \ldots, y_m). \qquad \square$$

It follows that existential formulas are upward persistent.

The converse of this proposition, that any downward persistent formula is logically universal, is also true but beyond the scope of this book. We know that the family of universal formulas is closed under conjunction and disjunction, so it must be the case that the family of downward persistent formulas is also closed under conjunction and disjunction. To see this for disjunction, let us suppose that ϕ and ψ are downward persistent and show that $\phi \vee \psi$ is also downward persistent. That is, suppose that $\mathcal{A} \subseteq \mathcal{B}$ and $\mathcal{B} \models \phi \vee \psi$. Then without loss of generality, $\mathcal{B} \models \phi$. Since ϕ is downward persistent, $\mathcal{A} \models \phi$, and it follows that $\mathcal{A} \models \phi \vee \psi$.

We will consider some other notions of persistence in the next section, including examples involving universal-existential formulas.

Exercise for Section 3.3

Exercise 4.3.1. Show that the class of equivalence structures with exactly three elements in each equivalence class is not downward persistent and therefore cannot have a universal axiomatization.

4.4.   Models, Theories, and Axioms

We begin with a useful definition.

Definition 4.4.1.

(1) Given a theory Γ in a fixed language $\mathcal{L}$, we let $\mathrm{Mod}(\Gamma)$ be the set of structures $\mathcal{A}$ such that $\mathcal{A} \models \gamma$ for all $\gamma \in \Gamma$.
(2) Given a class K of models in a fixed language $\mathcal{L}$, we let $Th(K)$ be the set of sentences γ such that $\mathcal{A} \models \gamma$ for all $\mathcal{A} \in K$.

For a single sentence γ, we let $\mathrm{Mod}(\gamma) = \mathrm{Mod}(\{\gamma\})$. Similarly, for a single structure $\mathcal{A}$, $\mathrm{Th}(\mathcal{A}) = \mathrm{Th}(\{\mathcal{A}\})$.

For example, in the language $\mathcal{L} = \{<\}$, consider the axioms of a partial ordering:

(a) $(\forall x)\neg(x < x)$.
(b) $(\forall x)(\forall y)\neg(x < y \land y < x)$.
(c) $(\forall x)(\forall y)(\forall z)((x < y \land y < z) \to x < z)$.

These axioms state that $<$ is irreflexive, antisymmetric, and transitive. If Γ_0 is the set of these three sentences, then $\mathrm{Mod}(\Gamma)$ is the class of partial orderings. If we add the axiom of trichotomy

(d) $(\forall x)(\forall y)(x < y \lor x = y \lor y < x)$

then the resulting set of four axioms define the class of linear orderings. Note that all of these axioms are universal, so that any substructure of a linear ordering is also a linear ordering.

Definition 4.4.2. A linear ordering $(A, <)$ is a *well-ordering* if every non-empty subset B of A has a least element b, that is, an element b such that $b \leq x$ for all $x \in B$. We note that this is not a first-order definition.

It is easy to see that $K \subseteq \mathrm{Mod}(\mathrm{Th}(K))$ for any class K of models and that $\Gamma \subseteq \mathrm{Th}(\mathrm{Mod}(\Gamma))$ for any Γ. However, $\mathrm{Th}(K)$ is always closed under implication, so Γ may be a proper subset of $\mathrm{Th}(\mathrm{Mod}(\Gamma))$. For example, if $\Gamma = \emptyset$, then $\mathrm{Mod}(\Gamma)$ consists of all possible structures for the given language and $\mathrm{Th}(\mathrm{Mod}(\Gamma))$ consists of all logically valid sentences.

A class K of models is said to be *axiomatizable* if $K = \mathrm{Mod}(\Gamma)$ for some set Γ of sentences. K is said to be *finitely axiomatizable* if $K = \mathrm{Mod}(\Gamma)$ for a finite set Γ; equivalently $K = \mathrm{Mod}(\gamma)$ for some single sentence γ. If K is not axiomatizable, then clearly $K \neq \mathrm{Mod}(\mathrm{Th}(K))$.

For a finite language $\mathcal{L}$, any finite structure $\mathcal{A}$ is finitely axiomatizable. For example, let $\mathcal{A} = (\{1,2\}, <)$ where $1 < 2$. Then the axiom for $\mathcal{A}$ is the following:

$$(\exists x)(\exists y)((x \neq y \;\wedge\; (\forall z)(x = z \;\vee\; y = z) \;\wedge\; x < y).$$

In general, the axiom for a finite structure $\mathcal{A}$ just states the existence of the distinct elements and defines the constants, relations and functions on $\mathcal{A}$ for those elements.

The language of pure equality provides a very interesting illustration of these notions, as well as the use of compactness. Recall from Section 4.2 the sentences $\mathcal{E}^{\geq n}$ stating that there are at least n distinct elements and the theory Γ_∞. Let $INF = \mathrm{Mod}(\Gamma_\infty)$ be the set of infinite structures. Proposition 4.2.6 showed that INF is not finitely axiomatizable.

Now consider the complementary set FIN of finite models.

Proposition 4.4.3. *FIN is not axiomatizable.*

Proof. Suppose by way of contradiction that there is a set Δ of sentences such that $FIN = \mathrm{Mod}(\Delta)$. Then $\Delta \cup \Gamma_\infty$ is inconsistent. So by compactness, there is a finite subset $\Gamma_n = \{\gamma_1, \ldots, \gamma_n\}$ such that $\Delta \cup \Gamma_n$ is inconsistent. But the structure $\{1, 2, \ldots, n\}$ satisfies Γ_n and it is finite, so it must satisfy Δ. This contradiction proves the result. $\qquad\square$

Notions of persistence may be framed in terms of classes of models. That is, a class K of models is downward persistent if $\mathcal{A} \subseteq \mathcal{B}$ and $\mathcal{B} \in K$ implies $\mathcal{A} \in K$. Then Proposition 4.3.21 implies that if $K = \text{Mod}(\Gamma)$ for a set Γ of universal formulas, then K is downward persistent.

Here we will examine some notions of persistence associated with *products* of structures as well as unions of chains of structures.

If $\mathcal{A}$ and $\mathcal{B}$ are two structures over the same language, with universes A and B, respectively, then $\mathcal{A} \times \mathcal{B}$ is the structure $\mathcal{C}$ with universe $A \times B$ defined as follows. For each constant symbol c, $c^C = (c^A, c^B)$. For each n-ary function symbol f, $f^C((a_1, b_1), \ldots, (a_n, b_n)) = (f^A(a_1, \ldots, a_n), f^B(b_1, \ldots, b_n))$. For each n-ary relation symbol R,

$$R^C((a_1, b_1), \ldots, (a_n, b_n)) \iff R^A(a_1, \ldots, a_n) \wedge R^B(b_1, \ldots, b_n).$$

We say that a formula ϕ is *product persistent* if whenever $\mathcal{A} \models \phi$ and $\mathcal{B} \models \phi$, then $\mathcal{A} \times \mathcal{B} \models \phi$.

For example, the group axioms given above are all product persistent, as the product of two groups is a group. The axioms for a commutative ring with unity in the language $(+, \times, 0, 1)$ are also product persistent. However, consider the additional axiom for being an integral domain, that there are no zero-divisors.

$$(\forall x)(\forall y)(x \times y = 0 \implies (x = 0 \ \vee \ y = 0)).$$

Example 4.4.4. $\mathbb{Z}_2$ and $\mathbb{Z}_3$ are both integral domains, but the product $\mathbb{Z}_2 \times \mathbb{Z}_3$ has zero divisors $(0, 1) \times (1, 0) = (0, 0)$. Hence, the statement that there are no zero-divisors is not product persistent.

Example 4.4.5. The sentence $\mathcal{E}^{\geq 2}$ is product persistent, since if both $\mathcal{A}$ and $\mathcal{B}$ have at least two elements, then $\mathcal{A} \times \mathcal{B}$ has at least 4 elements, so it satisfies $\mathcal{E}^{\geq 2}$. The formula expressing that there are *exactly* two elements is not product persistent.

Definition 4.4.6. The family of *Horn formulas* is the smallest family of formulas generated as follows. For all atomic formulas $\rho_1, \ldots, \rho_n, \theta$, the formulas $\neg(\rho_1 \wedge \rho_2 \wedge \cdots \wedge \rho_n)$ and $(\rho_1 \wedge \rho_2 \wedge \cdots \wedge \rho_n) \to \theta$ are Horn formulas. If ϕ and ψ are Horn formulas, then the conjunction $\phi \wedge \psi$ is a Horn formula and both $(\exists x)\phi$ and $(\forall x)\phi$ are Horn formulas.

We note that the statement about the non-existence of zero-divisors given above is *not* a Horn formula.

Proposition 4.4.7. *Any Horn formula ϕ is product persistent.*

The full proof is omitted here. The idea of the proof may be seen in the following.

Example 4.4.8. The property of transitivity of a relation R is product persistent. To see this, assume that structures $(\mathcal{A}, R^{\mathcal{A}})$ and $(\mathcal{B}, R^{\mathcal{B}})$ are transitive and suppose that $\mathcal{A} \times \mathcal{B} \models R((a_1, b_1), (a_2, b_2)) \wedge R((a_2, b_2), (a_3, b_3))$. By the definition of products, $\mathcal{A} \models R(a_1, a_2) \wedge R(a_2, a_3)$ and $\mathcal{B} \models R(b_1, b_2) \wedge R(b_2, b_3)$. Since both $\mathcal{A}$ and $\mathcal{B}$ are transitive, $\mathcal{A} \models R(a_1, a_3)$ and $\mathcal{B} \models R(b_1, b_3)$. Thus we have $\mathcal{A} \times \mathcal{B} \models R((a_1, b_1), (a_3, b_3))$. This shows that $\mathcal{A} \times \mathcal{B}$ is transitive.

Definition 4.4.9. A sequence of structures $\mathcal{A}_0 \subseteq \mathcal{A}_1 \subseteq \cdots$ is said to be a *chain*. The union $\mathcal{C}$ of such a chain has universe $\bigcup_i A_i$, where each structure $\mathcal{A}_i$ has universe A_i, and is defined as follows. For each constant symbol c, $c^{\mathcal{C}} = c^{\mathcal{A}_0}$. For each n-ary function symbol and for elements $a_1, \ldots, a_n$, let k be the least such that each $a_1, \ldots, a_n \in A_k$ and let $f^{\mathcal{C}}(a_1, \ldots, a_n) = f^{\mathcal{A}_k}(a_1, \ldots, a_n)$. For each n-ary relation symbol and for elements $a_1, \ldots, a_n$, let k be the least such that each $a_1, \ldots, a_n \in A_k$ and let $R^{\mathcal{C}}(a_1, \ldots, a_n) \iff R^{\mathcal{A}_k}(a_1, \ldots, a_n)$.

Proposition 4.4.10. *If $\mathcal{C}$ is the union of a chain $\mathcal{A}_0 \subseteq \mathcal{A}_1 \subseteq \cdots$, then for each k, $\mathcal{A}_k \subseteq \mathcal{C}$.*

Unions of chains preserve formulas with two levels of quantification. As an example, the statement that an ordering has no greatest element can be written as

$$(\forall y)(\exists x)(y < x)$$

and is therefore seen to be universal-existential.

Theorem 4.4.11. *For any $\forall\exists$ formula ϕ and any chain $\mathcal{A}_0 \subseteq \mathcal{A}_1 \subseteq \cdots$, if $\mathcal{A}_k \models \phi$ for all k, then the union $\mathcal{C}$ of the chain also satisfies ϕ.*

Proof. Let ϕ have the form $(\forall y_1)\cdots(\forall y_m)\theta(y_1,\ldots,y_m)$, where θ is existential, and suppose that $\mathcal{C}$ is the union of a chain $\mathcal{A}_0 \subseteq \mathcal{A}_1 \subseteq \cdots$ such that $\mathcal{A}_k \models \phi$ for each k. Let $c_1,\ldots,c_m \in C$ be given and take k large enough so that each $c_i \in \mathcal{A}_k$. Since $\mathcal{A}_k \models \phi$, it follows that $\mathcal{A}_k \models \theta(c_1,\ldots,c_m)$. Now θ is an existential formula and $\mathcal{A}_k \subseteq \mathcal{C}$, hence it follows from the upward persistence of existential formulas (a consequence of Proposition 4.3.21) that $\mathcal{C} \models \theta(c_1,\ldots,c_m)$. Since $c_1,\ldots,c_m$ are arbitrary, it follows that $\mathcal{C} \models (\forall y_1)\cdots(\forall y_m)\theta(y_1,\ldots,y_m)$, as desired.   □

Now we can see that the negation of the sentence above, which states that there *is* a greatest element in a linear ordering, cannot be logically $\forall\exists$, because the union of the chain $\mathcal{A}_k = [0,k]$ (under the usual ordering on the real numbers) would have no greatest element, whereas each $[0,k]$ does have a greatest element.

Exercises for Section 4.4

Exercise 4.4.1. Show the following:

(a) $K \subseteq \mathrm{Mod}(\mathrm{Th}(K))$ for any class K of models;
(b) $\Gamma \subseteq \mathrm{Th}(\mathrm{Mod}(\Gamma))$ for any set Γ of sentences.

Exercise 4.4.2. Show that $\mathrm{Mod}(\mathrm{Th}(K)) = K$ for any axiomatizable class K of models.

Exercise 4.4.3. Show that if Γ is a deductively closed theory, then $\text{Th}(\text{Mod}(\Gamma)) = \Gamma$.

Exercise 4.4.4. Write a sentence γ such that $\mathcal{A} = \text{Mod}(\gamma)$ where $\mathcal{A} = (\{0, 1\}, +)$ with addition mod 2.

Exercise 4.4.5. Show that the family of product persistent formulas is closed under conjunctions.

Exercise 4.4.6. Show that the class of well-orderings is not closed under unions of chains and therefore cannot have a $\forall\exists$ axiomatization.

Exercise 4.4.7. Show that the class of graphs of infinite degree is not finitely axiomatizable and that the class of graphs of finite degree is not axiomatizable. An unordered graph $G = (V, E)$ consists of a set V of elements (called vertices) and a binary edge relation E. The *degree* of a vertex v is the cardinality of $\{u : uEv\}$. G has finite degree k if k is the maximum of the degrees of the vertices in V. If no such finite maximum exists, then G is said to have infinite degree.

Exercise 4.4.8. Show by induction on Horn formulas that every Horn formula is product persistent.

Exercise 4.4.9. Show that if $\mathcal{C}$ is the union of a chain $\mathcal{A}_0 \subseteq \mathcal{A}_1 \subseteq \cdots$, then for each k, $\mathcal{A}_k \subseteq \mathcal{C}$.

4.5. Categoricity and Quantifier Elimination

One important tool for showing that a theory Γ is complete, i.e., that for every sentence ϕ in the language of Γ, either $\Gamma \vdash \phi$ or $\Gamma \vdash \neg\phi$, is the notion of quantifier elimination.

Definition 4.5.1. A first-order theory Γ is said to have *quantifier elimination* if, for any formula ϕ, there is a quantifier-free formula θ such that $\Gamma \vdash \phi \leftrightarrow \theta$. Here we include the quantifier-free sentences $\top$, which denotes a formula θ provable from Γ, and $\bot$, which denotes a formula θ such that $\Gamma \vdash \neg\theta$.

Note that in a relational language (with no function or constant symbols) a quantifier-free sentence must be empty, so that the process of quantifier elimination must end with a formula θ such that either $\Gamma \vdash \theta$ or $\Gamma \vdash \neg\theta$. Thus a relational theory with quantifier elimination is complete. If there is an algorithm to produce from any sentence ϕ a quantifier-free θ such that $\Gamma \vdash \phi \leftrightarrow \theta$, then this algorithm also decides whether $\Gamma \vdash \phi$ or $\Gamma \vdash \neg\phi$. Structures with functions and constants are a bit more complicated.

The theory of infinity, where $\Gamma_\infty = \{\mathcal{E}^{\geq 1}, \mathcal{E}^{\geq 2}, \dots\}$ as above, is a natural example of a theory which satisfies quantifier elimination.

Theorem 4.5.2. *The theory of infinity satisfies quantifier elimination, and there is an algorithm which produces an equivalent quantifier-free formula from any given formula.*

Proof. In the language of equality, the atomic formulas have the form $u = v$, where u and v are two variables, possibly identical. Now by the Disjunctive Normal Form Theorem, any quantifier-free formula $\theta(x_1, \dots, x_n)$ is equivalent to a disjunction $C_1 \vee \cdots \vee C_k$ of conjuncts $C_1, \dots, C_k$ where each C_i is a conjunct of literals, either of the form $u = v$ or the form $u \neq v$.

The result is proved by induction on the rank of a formula. For the base case, any atomic formula is quantifier-free already. For the connectives $\neg, \vee, \wedge$, if two formulas ϕ_1 and ϕ_2 are logically equivalent to quantifier-free formulas θ_1 and θ_2, respectively (under Γ_∞), then $\neg\phi_1$ is logically equivalent to $\neg\theta_1$, $\phi_1 \vee \phi_2$ is logically equivalent to $\theta_1 \vee \theta_2$ and $\phi_1 \wedge \phi_2$ is logically equivalent to $\theta_1 \wedge \theta_2$. But these formula $(\neg\theta_1, \theta_1 \vee \theta_2, \theta_1 \wedge \theta_1)$ are all quantifier-free.

Since $(\forall x)\theta$ is logically equivalent to $\neg(\exists x)\neg\theta$, it suffices to show that if $\theta(x, x_1, \ldots, x_n)$ is quantifier-free, then there is a quantifier-free formula ψ such that

$$\Gamma_\infty \vdash (\exists x)\theta(x, x_1, \ldots, x_n) \leftrightarrow \psi(x_1, \ldots, x_n).$$

Now for any disjunction $C_1 \vee \cdots \vee C_k$, $(\exists x)(C_1 \vee \cdots \vee C_k)$ is logically equivalent to $(\exists x)C_1 \vee (\exists x)C_2 \vee \cdots \vee (\exists x)C_k$, thus we may assume without loss of generality that θ is a conjunct of literals.

There are three cases to consider.

Case 1: One of the literals has the form $x \neq x$. In this case, $(\exists x)\theta$ is simply false.

Case 2: One of the literals has the form $x = x_i$. In this case we can modify the formula θ by replacing every occurrence of x with an occurrence of x_i to obtain the desired quantifier-free formula $\psi = \theta(x_i, x_1, \ldots, x_n)$ which is equivalent to $(\exists x)\theta$.

Case 3: Each of the literals has the form $x \neq x_i$. Then the sentence $\mathcal{E}^{\geq k+1}$ from Γ implies the existence of an element x which is different from each of $x_1, \ldots, x_k$. Thus we can modify θ by eliminating each literal in which x occurs to obtain the desired quantifier-free formula ψ.

Once we have eliminated all the quantifiers from a sentence φ, we have a quantifier-free formula with no variables, which must be either $\top$ or $\bot$, and this tells us whether φ is in the theory or not. $\qquad\square$

Let us observe that, for each infinite cardinal κ, there is a unique model, up to isomorphism, of the theory of infinity with cardinality κ. This leads to the following general notion.

Definition 4.5.3. A theory Γ is κ-*categorical* for some infinite cardinal κ if and only if every two models of Γ of cardinality κ are isomorphic.

The key result here is that theories that are categorical in some power and have only infinite models are also complete.

Theorem 4.5.4. (Łoś-Vaught Test). *Suppose that Γ is a theory of cardinality κ with no finite models. If Γ is λ-categorical for some (infinite) $\lambda \geq \kappa$, then Γ is complete.*

Proof. Suppose by way of contradiction that Γ is not complete, but is λ-categorical for some $\lambda \geq \kappa$. Then there is a sentence φ in the language of Γ such that neither $\Gamma \vdash \varphi$ nor $\Gamma \vdash \neg\varphi$. Let $\Gamma_1 = \Gamma \cup \{\varphi\}$ and $\Gamma_2 = \Gamma \cup \{\neg\varphi\}$. Since Γ is consistent, so are Γ_1 and Γ_2. By the Completeness Theorem, each of these theories has a model. Since both of these models are also models of Γ, by hypothesis, they must be infinite. Therefore by the Löwenheim–Skolem Theorem, they each have models of cardinality λ. Let $\mathcal{N}_1$ and $\mathcal{N}_2$ be models of Γ_1 and Γ_2 of cardinality λ. Both $\mathcal{N}_1$ and $\mathcal{N}_2$ are models of Γ of power λ, so they are isomorphic. By Theorem 4.3.4, $\mathcal{N}_1$ and $\mathcal{N}_2$ must be elementarily equivalent. However $\mathcal{N}_1$ is a model of φ while $\mathcal{N}_2$ is a model of $\neg\varphi$. This contradiction proves that Γ is complete.   □

Note that the theory of infinity is $\aleph_0$-categorical. Another example of an $\aleph_0$-categorical theory is the theory of dense linear orders without endpoints.

Consider the following sentences that state the existence of endpoints of our linear order, REnd for right endpoint and LEnd for left endpoint:

$$(\exists x)(\forall y)(x \leq y),$$
$$(\exists x)(\forall y)(y \leq x).$$

The theory of *dense linear orders without endpoints*, denoted *DLOWE*, is the theory of dense linear orderings with the addition of the negations of REnd and LEnd. We will show that this theory is $\aleph_0$-categorical.

One model of the theory is quite familiar, since the rationals with the usual order is a model:

$$\langle \mathbb{Q}, \leq \rangle \models DLOWE.$$

We will establish the completeness of $DLOWE$ by proving a series of results.

Theorem 4.5.5. *Any nonempty model of DLOWE is infinite.*

Proof. Left to the reader. □

Theorem 4.5.6. *Any two nontrivial countable models of DLOWE are isomorphic.*

Proof. Suppose that $\mathcal{A} = \langle A, \leq_A \rangle$ and $\mathcal{B} = \langle B, \leq_B \rangle$ are two nonempty countable models of $DLOWE$. Suppose that $\langle a_i : i < \omega \rangle$ and $\langle b_i : i < \omega \rangle$ are enumerations of A and B, respectively. We define an isomorphism $h : A \to B$ by defining h in a sequence of stages.

At stage 0 of our construction, set $h(a_0) = b_0$. Suppose at stage $m > 0$, h has been defined on $\{a_{i_1}, \ldots, a_{i_k}\}$ where the elements of A are listed in increasing order under the relation $\leq_A$. Further suppose that we denote by b_{r_j} the value of $h(a_{i_j})$. Then since by hypothesis h is an isomorphism on the points on which it is defined, the elements $b_{r_1}, \ldots, b_{r_k}$ are listed in increasing order under the relation $\leq_B$.

At stage $m = 2n$ of our construction, we ensure that a_n is in the domain of h. If $h(a_n)$ has already been defined at an earlier stage, then there is nothing to do at stage $m = 2n$. Otherwise, either (i) $a_n <_A a_{i_1}$, (ii) $a_{i_k} <_A a_n$, (iii) or for some ℓ, $a_{i_\ell} <_A a_n <_A a_{i_{\ell+1}}$. Choose b_r as the element of B of least index in the enumeration $\langle b_i : i < \omega \rangle$ that has the same relationship to $b_{r_1}, \ldots, b_{r_k}$ that a_n has to $a_{i_1}, \ldots, a_{i_k}$. It is possible to choose such a b_r in (i) the first case because B has no left endpoint, (ii) the second case because B has no right endpoint, and (iii) the last case because the order is dense. Extend h to a_n by setting $h(a_n) = b_r$.

At stage $m = 2n + 1$ of our construction, we ensure that the point b_n is in the range of h. As above, if b_n is in the range of h, then there is nothing to do at stage $m = 2n+1$. Otherwise, it has a unique position relative to $b_{r_1}, \ldots, b_{r_k}$. As above, either it is a left endpoint, a right endpoint, or it lies strictly between b_{r_ℓ} and $b_{r_{\ell+1}}$ for some ℓ. As in the previous case, choose a_r as the element of A of least index in the enumeration of $\langle a_i : i < \omega \rangle$ which has the same relationship to $a_{i_1}, \ldots, a_{i_k}$ as b_n has to $b_{r_1}, \ldots, b_{r_k}$, and extend h to a_r by setting $h(a_r) = b_n$.

This completes the recursive definition of h. One can readily prove by induction that the domain of h is A, the range of h is B, and that h is an isomorphism. ☐

We can now apply the Łoś-Vaught Test to conclude:

Corollary 4.5.7. *The theory DLOWE is complete.*

Exercises for Section 4.5

Exercise 4.5.1. Show that any linear order with no right endpoint must be infinite. Show that any dense linear order with at least two points must be infinite.

Exercise 4.5.2. Show that the theory *DLOWE* has quantifier elimination.

4.6.   Examples of Theories and Structures

The main goal of this section is to provide a complete set of axioms for the structure $\mathbb{R}$ of the real numbers, in the language which has function symbols $+$, $\times$, constant symbols 0 and 1, and one two-place relation, $\leq$. For simplicity, we omit symbols for the additive and multiplicative inverse. Then the structure of the real numbers is $\mathbb{R} = (R, +^{\mathbb{R}}, \times^{\mathbb{R}}, 0^{\mathbb{R}}, 1^{\mathbb{R}}, \leq^{\mathbb{R}})$. We will generally omit the superscripts $\mathbb{R}$ when we have the standard operations on real numbers.

Recall from Section 4.2 that $(\mathbb{R}, +, 0)$ is a commutative group. Adding the multiplication operation, we obtain the axioms for *rings* and *fields*.

Definition 4.6.1. A structure $\mathcal{A} = (A, +, \times, 0, 1)$ is a *ring* if $(A, +, 0)$ is a commutative group (satisfying axioms (1)–(4) from Definition 4.2.10) and $\mathcal{A}$ satisfies the following additional axioms:

(5) Associative Law: $(\forall x)(\forall y)(\forall z)(x \times (y \times z) = (x \times y) \times z)$;
(6) Identity Law: $(\forall x)(x \times 1 = x)$;
(7) Distributive Law: $(\forall x)(\forall y)(\forall z)(x \times (y + z) = (x \times y) + (x \times z))$.
   The ring $\mathcal{A}$ is *commutative* if it also satisfies;
(8) Commutative Law: $(\forall x, y)(x \times y = y \times x)$.

Example 4.6.2. $\mathbb{Z}$ and $\mathbb{Q}$ with the operations of $+$ and $\times$ are commutative rings. Let $\mathcal{M}_2 = (M, +, *, O, I)$ be the set of two by two matrices with real entries, with the usual matrix addition and multiplication, and the usual identity matrices. This is a ring, but is not commutative.

The theory of the real numbers includes axioms stating that $\mathbb{R}$ is an ordered field.

Definition 4.6.3. A structure $\mathcal{A} = (A, +, \times, 0, 1)$ is a *field* if $(A, +, \times, 0)$ is a commutative ring and $\mathcal{A}$ satisfies the following additional axiom:

(9) Inverse Law: $(\forall x \neq 0)(\exists y)(x \times y = 1)$.

Example 4.6.4. $\mathbb{Q}$ and $\mathbb{R}$ are fields but $\mathbb{Z}$ is not a field. The structure $\mathbb{Z}_n = (\{0, 1, \ldots, n-1\}, +, \times, 0, 1)$ with addition and multiplication taken modulo n, is always a commutative ring and is a field if and only if n is a prime number.

Given these laws, we may define the notions of subtraction and division.

Definition 4.6.5.

(1) *Subtraction.* For any real numbers x and y, $x - y$ is the unique real number z such that $y + z = x$.
(2) *Division.* For any real number $y \neq 0$ and any real number x, x/y is the unique real number z such that $y \times z = x$.

It is left as an exercise to show that these operations are well-defined.

Finally, we consider the ordering on the real numbers.

Definition 4.6.6. An ordered field $\mathcal{A} = (A, +, \times, 0, 1, \leq)$ is a field equipped with a linear ordering $\leq$ such that

(10) $(\forall x)(\forall y)(\forall z)(x \leq y \rightarrow (x + z \leq y + z))$;
(11) $(\forall x)(\forall y)(\forall z > 0)(x \leq y \rightarrow (x \times z \leq y \times z))$.

Example 4.6.7. $\mathbb{Q}$ and $\mathbb{R}$ are ordered fields. Finite fields, such as $\mathbb{Z}_2 = \{0, 1\}$ cannot be ordered.

We can now describe a complete set of axioms for $\mathbb{R}$. Recall that a polynomial $p(x) = a_0 + a_1 x + \cdots + a_n x^n$ with $a_n \neq 0$ is said to have degree n and that z is a *zero* for the function p if $p(z) = 0$. An element y is said to be a *square root* of element x if $y \times y = x$. This begins with defining two axiom schemas.

Definition 4.6.8.

(12) $(\forall x)(x > 0 \rightarrow (\exists y)(y \times y = x))$;
$(13)_n$ $(\forall a_0)(\forall a_1) \cdots (\forall a_n)(a_n \neq 0 \rightarrow (\exists x)(a_0 + a_1 x + \cdots + a_n x^n = 0))$.

Definition 4.6.9. A real closed field $(K, +, \times, 0, 1, \leq)$ is an ordered field such that every positive element x has a square root and such that every polynomial $p(x)$ of odd degree has a zero. That is, K satisfies axiom (12) and satisfies the axiom scheme $(13)_n$ for each odd natural number n.

Alfred Tarski proved that the axioms for a real closed field are complete and the Tarski–Seidenberg theorem shows that it also has quantifier elimination. See [33].

Theorem 4.6.10 (Tarski). *The axioms for a real closed field are complete and the theory has quantifier elimination.*

We observe that $\mathbb{R}$ is not the unique real closed field. The set of *algebraic numbers*, that is, zeros of polynomials with rational coefficients, is in fact a countable real closed field.

There are two other equivalent ways to state the real closure of a field, both related to the ordering.

Definition 4.6.11. The definable Least Upper Bound Principle consists of the following scheme (stated informally):

For any formula $\varphi(x)$ with one free variable (as well as parameters), if there is an upper bound b such that $(\forall x)\varphi(x) \to x \le b$, then there is a least upper bound, that is, a least upper bound for $\{x : \varphi(x)\}$.

Definition 4.6.12. The Intermediate Value Principle consists of the following scheme (stated informally):

For any continuous function f of a single variable which is definable from parameters, any $a < b$, and any v between $f(a)$ and $f(b)$, there is some $c \in [a, b]$ such that $f(c) = v$.

Proposition 4.6.13. *The following conditions are equivalent:*

(a) *The Real Closure axioms* (12) *and* $(13)_n$ *for odd* $n \in \mathbb{N}$.
(b) *The definable Least Upper Bound Principle.*
(c) *The definable Intermediate Value Principle.*

Proof. We will just examine the implications from (b) to (a) and from (b) to (c).

Suppose K is an ordered field in which the definable Least Upper Bound Principle holds. Let a be a positive element of K and let $S = \{x : x \times x \le a\}$. Note that this set is definable. Let $x \in S$. If $x \ge 1$, then $x \le x \times x \le a$, so S has upper bound

$a + 1$. Thus by the LUB Principle S has a least upper bound b. We claim that $b \times b = a$. The proof is by way of contradiction. Suppose first that $b \times b < a$. Then, by the continuity of multiplication, there is some positive ϵ, $(b+\epsilon)(b+\epsilon) < a$. But this means that $b + \epsilon \in S$, and $b < b + \epsilon$, contradicting the assumption that b is an upper bound for S. Suppose next that $b \times b > a$. Then there is some positive ϵ so that $(b-\epsilon)(b-\epsilon) > a$. But this means that $b - \epsilon$ is an upper bound for S, and $b > b - \epsilon$, contradicting the assumption that b is the least upper bound for S.

The proof that polynomials of odd degree have zeros is left as an exercise.

Next suppose that $f : K \to K$ is continuous, let $a < b$, and assume without loss of generality, that $f(a) < v < f(b)$. The argument here is a bit more complicated than the argument for having square roots, since the function $f(x) = x^2$ is increasing for $x > 0$ but here we cannot assume that f is increasing. So we let

$$S = \{x \leq b : (\forall y)(a \leq y < x \to f(y) < v)\},$$

and let c be the least upper bound of S. Note that as b is an upper bound of S, we have $c \leq b$. We claim that $f(c) = v$. The proof is by way of contradiction. Suppose first that $f(c) < v$, so that $f(c) < v < f(b)$ and hence $c < b$. Then, by continuity, there is some positive $\epsilon < b - c$ such that $f(x) < v$ for all x with $c \leq x < c+\epsilon$. But this means that $c + \epsilon \in S$, contradicting the assumption that c is an upper bound for S. Suppose next that $f(c) > v$. Then there is some positive ϵ so that $f(x) > v$ for all x with $c - \epsilon \leq x < c$. But this means that $c - \epsilon$ is an upper bound for S, contradicting the assumption that c is the least upper bound for S. □

Exercises for Section 4.6

Exercise 4.6.1. Show that the axioms for a commutative group are not complete. *Hint*: Make use of a difference between $\mathbb{Z}$ and $\mathbb{Q}$.

Exercise 4.6.2. Show that the field $\mathbb{Z}_3 = \{0, 1, 2\}$ with addition and multiplication modulo 3 cannot be ordered. *Hint*: Start with the assumption that either $0 < 1$ or $1 < 0$.

Exercise 4.6.3. Show that the definable Least Upper Bound Principle implies that every polynomial of odd degree has zeros.

Chapter 5

Boolean Algebras

5.1. Properties and Examples of Boolean Algebras

In this section, we present the properties of Boolean algebras and study some examples, such as the family of subsets of a given set, and the *Lindenbaum algebra* arising from propositional as well as predicate logic. There are many formulations for the definition of a Boolean algebra. This subject was originated by George Boole [3]. We follow the approach of Stoll [31]. For further study, see also [28].

Definition 5.1.1. A *Boolean algebra* $\mathcal{B} = (B, \vee, \wedge, -, 0^B, 1^B)$ is a set together with operations $\vee$ (*join*), $\wedge$ (*meet*), $-$ (*complement*) and *identity elements* 0^B and 1^B which satisfy the axioms in the theory of Boolean algebras given below (hereafter we will write 0^B and 1^B as 0 and 1, respectively). The set B is called the *universe* of the Boolean algebra $\mathcal{B}$.

Definition 5.1.2. Axioms for the theory of Boolean algebras:

(1) Commutative Laws: $(\forall a)(\forall b)(a \vee b = b \vee a)$ and $(\forall a)(\forall b)(a \wedge b = b \wedge a)$;

(2) Associative Laws: $(\forall a)(\forall b)(\forall c)(a \vee (b \vee c) = (a \vee b) \vee c)$ and $(\forall a)(\forall b)(\forall c)(a \wedge (b \wedge c) = (a \wedge b) \wedge c)$;

(3) Distributive Laws: $(\forall a)(\forall b)(\forall c)(a \wedge (b \vee c) = (a \wedge b) \vee (a \wedge c))$
and $(\forall a)(\forall b)(\forall c)(a \vee (b \wedge c) = (a \vee b) \wedge (a \vee c))$;
(4) Complement Laws: $(\forall a)(a \wedge -a = 0)$ and $(\forall a)(a \vee -a = 1)$;
(5) Identity Laws: $(\forall a)(a \vee 0 = a)$ and $(\forall a)(a \wedge 1 = a)$.

Here are some important examples.

Example 5.1.3. For any set U, let the power set $\mathcal{P}(U)$ be the set of subsets of U. Then $(\mathcal{P}(U), \cup, \cap, {}^c, \emptyset, U)$ is a Boolean algebra, where A^c is the complement $U \setminus A$. We verified in Section 2.2 that $\mathcal{P}(U)$ satisfies the commutative and associative properties for union, as well as one of the distributive properties.

Example 5.1.4. We recall the notions of truth interpretations and truth functions from Chapter 1. For a finite set $\{A_0, \ldots, A_{n-1}\}$ of propositional variables, we can let $\mathcal{I}_n$ be the set of interpretations of $\{A_0, \ldots, A_{n-1}\}$. Then the set $\mathcal{F}_n$ of truth functions $F : \mathcal{I}_n \to \{0, 1\}$ is a Boolean algebra with the operations $F \vee G$, $F \wedge G$, and $-F$ defined as follows: $(F \vee G)(I) = \max\{F(I), G(I)\}$; $(F \wedge G)(I) = \min\{F(I), G(I)\}$; $-F(I) = 1 - F(I)$. Then $\mathcal{F}_n$ is isomorphic to the Boolean algebra $\mathcal{B}(\mathcal{L})(A_0, \ldots, A_{n-1})$ of propositions over the set $\{A_0, \ldots, A_{n-1}\}$ of propositional variables.

Recall that for any propositional sentence φ, there is a truth function TF_φ such that $\mathrm{TF}_\varphi(I) = I(\varphi)$. The following observations are not hard to show.

(1) $\mathrm{TF}_\varphi \vee \mathrm{TF}_\psi = \mathrm{TF}_{\varphi \vee \psi}$;
(2) $\mathrm{TF}_\varphi \wedge \mathrm{TF}_\psi = \mathrm{TF}_{\varphi \wedge \psi}$;
(3) $-\mathrm{TF}_\varphi = \mathrm{TF}_{\neg \varphi}$.

These are left as exercises.

This leads naturally to the notion of a Lindenbaum algebra for propositional (and also predicate) logic.

Definition 5.1.5. Fix a language $\mathcal{L}$ for propositional or predicate logic and recall the equivalence relation $\Longleftrightarrow$, of provable equivalence, defined so that $P \iff Q$ if and only if $\vdash P \leftrightarrow Q$. The *Lindenbaum algebra* $\mathcal{B}(\mathcal{L})$ is the set of equivalence classes of sentences of $\mathcal{L}$ equipped with the following interpretations:

(1) $0^{\mathcal{B}(\mathcal{L})} = [A_0 \wedge \neg A_0]$;
(2) $1^{\mathcal{B}(\mathcal{L})} = [A_0 \vee \neg A_0]$;
(3) $-[P] = [\neg P]$;
(4) $[P] \wedge [Q] = [P \wedge Q]$;
(5) $[P] \vee [Q] = [P \vee Q]$.

It needs to be checked that this definition does not depend on the choice of representatives of equivalence classes.

Proposition 5.1.6. *For any sentences* P_0, P_1, Q_0, Q_1, *if* $[P_0] = [P_1]$ *and* $[Q_0] = [Q_1]$, *then*

(1) $-[P_0] = -[P_1]$;
(2) $[P_0] \wedge [Q_0] = [P_1] \wedge [Q_1]$;
(3) $[P_0] \vee [Q_0] = [P_1] \vee [Q_1]$.

Proof. We will give the second proof, and leave the others as exercises. Suppose that $[P_0] = [P_1]$ and $[Q_0] = [Q_1]$. This means that $P_0 \iff P_1$ and $Q_0 \iff Q_1$. It follows that $P_0 \wedge Q_0 \iff P_1 \wedge Q_1$. Then $[P_0] \wedge [Q_0] = [P_0 \wedge Q_0] = [P_1 \wedge Q_1] = [P_1] \wedge [Q_1]$.
□

The *interval algebra* associated with a linear ordering provides another interesting Boolean algebra.

Example 5.1.7. Let $\mathcal{B}$ be the set of finite unions of left-open, right-closed intervals of the real line, with the standard union, intersection, and complement as the Boolean operations. Note that the left-open, right-closed intervals have one of the following forms, for some $a < b \in \mathbb{R}$: (1) $\emptyset$; (2) $\mathbb{R}$; (3) $(-\infty, b]$; (4) $(a, b]$; (5) (a, ∞). It is not hard to check that this family of sets is closed under the operations. It suffices to show that the intersection of any two such sets is another such set. A similar definition can be given for the set of finite unions of right-open, left closed sets.

Here are some important consequences of the Boolean algebra axioms for the identities and complements.

Proposition 5.1.8.

(6) *The identity elements* 0 *and* 1 *are the unique elements which satisfy item* (5).
(7) *Each element has a unique complement.*
(8) *For each element* a, $-(-a) = a$.
(9) $-0 = 1$ *and* $-1 = 0$.

Proof.

(6) Suppose that there are other elements $0'$ and $1'$ such that $a \vee 0' = a$ and $a \wedge 1' = a$ for all elements a. Then letting $a = 0$, we have $0 \vee 0' = 0$, but $0 \vee 0' = 0' \vee 0 = 0'$ by (1) and (5). Thus $0 = 0'$. The proof for 1 is similar.
(7) Suppose there are elements b and c such that $a \wedge b = 0 = a \wedge c$ and $a \vee b = 1 = a \vee c$. Then

$$b = 1 \wedge b = (a \vee c) \wedge b = (a \wedge b) \vee (c \wedge b) = 0 \vee (c \wedge b) = c \wedge b.$$

Similarly, $c = b \wedge c = c \wedge b$, so that $b = c$.
(8) Both a and $-(-a)$ are the inverse of $-a$, so it follows from (7) that $a = -(-a)$.
(9) Left as an exercise. $\square$

Here are some additional laws which follow from the definition of a Boolean algebra.

Proposition 5.1.9. *Any Boolean algebra satisfies the following:*

(10) *Idempotent Laws:* $(\forall a)\ a \vee a = a$ *and* $(\forall a)\ a \wedge a = a$;
(11) $(\forall a)\ a \wedge 0 = 0$ *and* $(\forall a)\ a \vee 1 = 1$;
(12) *Absorption Laws:* $(\forall a)(\forall b)\ a \vee (a \wedge b) = a$ *and* $(\forall a)(\forall b)\ a \wedge (a \vee b) = a$;
(13) *De Morgan's Laws:* $(\forall a)(\forall b) - (a \wedge b) = -a \vee -b$ *and* $(\forall a)(\forall b) - (a \vee b) = -a \wedge -b$.

Proof.

(10)
$$a = a \wedge 1 \qquad \text{by (5)}$$
$$= a \wedge (a \vee -a) \qquad \text{by (4)}$$
$$= (a \wedge a) \vee (a \wedge -a) \qquad \text{by (3)}$$
$$= (a \wedge a) \vee 0 \qquad \text{by (4)}$$
$$= a \wedge a \qquad \text{by (5)}.$$

The first part is similar.

(11) Left as an exercise.

(12) We will show that $-a$ is the inverse of $a \wedge (a \vee b)$, so that $a = a \wedge (a \vee b)$ by (7).

$$-a \wedge (a \wedge (a \vee b)) = (-a \wedge a) \wedge (a \vee b) \qquad \text{by (2)}$$
$$= 0 \wedge (a \vee b) \qquad \text{by (4)}$$
$$= 0 \qquad \text{by (11)}.$$

The other identity is left as an exercise.

(13) Left as an exercise. □

Exercises for Section 5.1

Exercise 5.1.1. Show that for all a in a Boolean algebra, $a \vee a = a$.

Exercise 5.1.2. Show that for all a in a Boolean algebra, $a \wedge 0 = 0$ and $a \vee 1 = 1$. *Hint*: Use part (10).

Exercise 5.1.3. Prove that, for all a, b, $a \vee (a \wedge b) = a$. *Hint*: Show that $-a$ is the inverse of $(a \vee (a \wedge b))$.

Exercise 5.1.4. Prove De Morgan's Laws from the axioms for Boolean algebras. *Hint*: Show that $a \wedge b$ is the inverse of $-a \vee -b$.

Exercise 5.1.5. Show that in the Boolean algebra F_n of truth functions,

(a) $\mathrm{TF}_\varphi \vee \mathrm{TF}_\psi = \mathrm{TF}_{\varphi \vee \psi}$;
(b) $\mathrm{TF}_\varphi \wedge \mathrm{TF}_\psi = \mathrm{TF}_{\varphi \wedge \psi}$;
(c) $-\mathrm{TF}_\varphi = \mathrm{TF}_{\neg \varphi}$.

Exercise 5.1.6. Show that the Boolean algebra $\mathcal{F}_n$ of truth functions over n propositional variables is isomorphic to the Lindenbaum algebra of propositional sentences $\mathcal{B}(\{A_1, \ldots, A_n\})$.

Exercise 5.1.7. Show that the Boolean algebra $\mathcal{P}(X)$ satisfies the identity laws.

Exercise 5.1.8. Show that the Boolean algebra $\mathcal{P}(X)$ satisfies the complement laws.

Exercise 5.1.9. Show that the Boolean algebra $\mathcal{P}(X)$ satisfies the first distributive law.

Exercise 5.1.10. Complete the proof of Lemma 5.1.6 by showing that $-[P_0] = -[P_1]$ and $[P_0] \vee [Q_0] = [P_1] \vee [Q_1]$.

5.2. The Partial Ordering on a Boolean Algebra

In this section, we discuss the partial ordering given by a Boolean algebra.

Definition 5.2.1. Let $\mathcal{B} = (B, \vee, \wedge, -^B, 0^B, 1^B)$ be a Boolean algebra. For elements $a, b \in B$, define $a \leq b$ if $a \wedge b = a$. If $a \leq b$ and $a \neq b$, we write $a < b$.

Proposition 5.2.2. *Let* $\mathcal{B} = (B, \vee, \wedge, -^B, 0^B, 1^B)$ *be a Boolean algebra. Then the relation* $a \leq b$ *is*

(1) *reflexive:* $a \leq a$;
(2) *antisymmetric: if* $a \leq b$ *and* $b \leq a$, *then* $a = b$;
(3) *transitive: if* $a \leq b$ *and* $b \leq c$, *then* $a \leq c$.

Proof. It follows from (10) that $a \wedge a = a$, so that $a \leq a$. The other parts are left to the exercises. $\qquad \square$

Proposition 5.2.3. *For any elements* a, b, c *in a Boolean algebra* $\mathcal{B}$,

(i) $a \vee b \leq c$ *if and only if* $a \leq c$ *and* $b \leq c$;
(ii) $c \leq a \wedge b$ *if and only if* $c \leq a$ *and* $c \leq b$.

Proof. (i) We have $a \wedge (a \vee b) = a$ by Absorption, so that $a \leq a \vee b$; $b \leq a \vee b$ follows by a similar argument. Now suppose that $a \vee b \leq c$. Then $a \leq c$ and $b \leq c$ by transitivity of the partial order.

Next suppose that $a \leq c$ and $b \leq c$. Then $a \wedge (a \vee c) = a$ and $b \wedge (b \vee c) = b$. Thus $(a \vee b) \wedge (a \vee b \vee c) = (a \wedge (a \vee b \vee c)) \vee (b \wedge (a \vee b \vee c)) = a \vee b$, again by Absorption. Hence $a \vee b \leq c$.

(ii) This is left as an exercise. $\square$

It is interesting to evaluate the ordering on the particular examples of Boolean algebras from the previous section.

Proposition 5.2.4.

(1) *On* $\mathcal{P}(U)$, *the ordering is given by the subset relation.*
(2) *On* $\mathcal{F}_n$, *the ordering is given by* $F \leq G$ *if and only if, for all truth interpretations* I, $F(I) \leq G(I)$.
(3) *On the Lindenbaum algebra* $\mathcal{B}(\mathcal{L})$ *of a language* $\mathcal{L}$, *the ordering is given by provable implication.*
(4) *On the Boolean algebra of finite unions of half-open intervals, the ordering is given by the subset relation.*

Proof. Here is the proof of part (3). Suppose first that $[P] \leq [Q]$. Then $[P] \wedge [Q] = [P]$, so that $P \wedge Q \Longleftrightarrow P$. Since $P \wedge Q \vdash Q$, it follows that $P \vdash Q$. For the converse, suppose that $P \vdash Q$. Then $P \vdash P \wedge Q$. But it is always true that $P \wedge Q \vdash P$. Therefore $P \wedge Q \Longleftrightarrow P$, so that $[P] \wedge [Q] = [P]$, as desired. $\square$

Definition 5.2.5. An element a of a Boolean algebra $\mathcal{B}$ is an *atom* if and only if $a \neq 0$ and for all elements b of the Boolean algebra, if $b \leq a$ then $b = 0$ or $b = a$.

In the Boolean algebra $\mathcal{P}(U)$, the atoms are just the singletons $\{u\}$ for each $u \in U$. In the Boolean algebra $\mathcal{F}_n$, the atoms are the functions F such that $F(I) = 1$ for exactly one interpretation. Thus in the corresponding Boolean algebra $\mathcal{B}(\mathcal{L})(A_0, \ldots, A_{n-1})$ the atoms have the form $[B_0 \wedge B_1 \wedge \cdots \wedge B_{n-1}]$, where each B_i is either A_i or $\neg A_i$.

Definition 5.2.6. Let $\mathcal{B}$ be a Boolean algebra $\mathcal{B}$.

(1) $\mathcal{B}$ is *atomless* if it has no atoms.
(2) $\mathcal{B}$ is *atomic* if for every $b \neq 0$ in B, there is an atom $a \leq b$.

Definition 5.2.7. A partial ordering $(X, \leq)$ is said to be *dense* if, for any $a < b$ in X, there exists $c \in X$ such that $a < c < b$. A Boolean algebra is dense if the associated partial ordering is dense.

Proposition 5.2.8. *A Boolean algebra $\mathcal{B}$ is dense if and only if it is atomless.*

Proof. Assume first that $\mathcal{B}$ is dense. Now let b be any element such that $0 < b$. Then there exists c with $0 < c < b$. Thus b is not an atom. Hence $\mathcal{B}$ is atomless.

Next suppose that $\mathcal{B}$ is atomless and let $a < b$. Then $a \wedge b = a$, by the definition of $\leq$. We claim that $0 < b \wedge -a$. To see this, suppose that $b \wedge -a = 0$. Then

$$b = b \wedge 1$$
$$= b \wedge (a \vee -a)$$
$$= (b \wedge a) \vee (b \wedge -a)$$
$$= a \vee 0$$
$$= a.$$

But this contradicts $a < b$. Thus $0 < b \wedge -a$. Since $\mathcal{B}$ is atomless, there is some c with $0 < c < b \wedge -a$. We claim that $a < a \vee c < b$. First we show that $a < a \vee c$. Certainly $a \leq a \vee c$ by

Proposition 5.2.3. Suppose by way of contradiction that $a = a \lor c$. Then by Exercise 5.2.2, $c \le a$. Since $c \le b \land -a$, it follows from Proposition 5.2.3 that $c \le a \land b \land -a = 0$, so that $c = 0$.

Next we show that $a \lor c < b$. Since $c < b \land -a$, we have $c \le b$ by Proposition 5.2.3, and therefore $a \lor c \le b$ since $a < b$ is assumed. Now suppose by way of contradiction that $b \le a \lor c$. Then

$$b \land -a \le (a \lor c) \land -a$$
$$= (a \land -a) \lor (c \land -a)$$
$$= 0 \lor (c \land -a)$$
$$\le c.$$

This contradicts the assumption that $c < b \land -a$.

Now we have $a < a \lor c < b$. Since $a < b$ was arbitrary, this shows that $\mathcal{B}$ is dense. □

It is easy to see that, for any set U, $\mathcal{P}(U)$ is atomic. The finite Lindenbaum algebra $\mathcal{B}(\mathcal{L})(A_0, \ldots, A_{n-1})$ is also atomic.

Proposition 5.2.9. *The propositional Boolean algebra $\mathcal{B}(\mathcal{L})$ over an infinite set $\{A_0, A_1, \ldots\}$ of variables is atomless.*

Proof. Let $[P] \ne 0$ in $\mathcal{B}(\mathcal{L})$. Then P is satisfiable. Now let A_n be some variable not occurring in P. Then $A_n \land P$ is also satisfiable and certainly $A_n \land P \vdash P$. However, it is not the case that $P \vdash A_n$, since for any I with $I(P) = 1$, we can modify I to J such that $J(A_n) = 0$ and still have $J(P) = 1$. Thus $0 < [P \land A_n] < [P]$. □

Example 5.2.10. Recall the Boolean algebra $\mathcal{B}$ consisting of finite unions of left-open, right-closed intervals of the real line, with the standard union, intersection, and complement as the Boolean operations. This can be shown to be atomless, as follows. Given a nonempty set $(a, b]$, we must have $a < b$ and then $(\frac{1}{2}(a + b), b]$ is a proper subset of $(a, b]$. For the infinite intervals, we have $(a + 1, \infty) \subsetneq (a, \infty)$ and $(-\infty, b - 1] \subsetneq (-\infty, b]$.

Exercises for Section 5.2

Exercise 5.2.1. Show that for any x, y in a Boolean algebra, $x \wedge y \leq x$.

Exercise 5.2.2. Show that for any x, y in a Boolean algebra, $x \leq y$ if and only if $x \vee y = y$.

Exercise 5.2.3. Show that the relation $x \leq y$ from Proposition 5.2.2 is antisymmetric and transitive.

Exercise 5.2.4. Show that in the Boolean algebra $\mathcal{P}(U)$ of subsets of U, $A \subseteq B \iff A \cap B = A$.

Exercise 5.2.5. Show that a is an atom of a Boolean algebra $\mathcal{B}$ if and only if, for all b and c, if $a = b \vee c$, then either $a = b$ or $a = c$.

Exercise 5.2.6. Show that for any set U, $\mathcal{P}(U)$ is atomic.

Exercise 5.2.7. Show that the finite Lindenbaum algebra $\mathcal{B}(\mathcal{L})(A_0, \ldots, A_{n-1})$ is atomic.

Exercise 5.2.8. Show that in the Boolean algebra $\mathcal{F}_n$, $F \leq G \iff \{I : F(I) = 1\} \subseteq \{I : G(I) = 1\}$, or equivalently, for all I, $F(I) \leq G(I)$.

Exercise 5.2.9. Verify that the family of finite unions of left-open, right-closed sets is closed under the Boolean operations.

Exercise 5.2.10. Show that the family of subsets of $\mathbb{N}$ which are either finite or cofinite is a Boolean algebra under the usual operations. Determine whether this Boolean algebra is atomic, or atomless, or neither.

5.3.   Filters and Ideals

Definition 5.3.1. Let $\mathcal{B} = (B, \vee, \wedge, -^B, 0^B, 1^B)$ be a Boolean algebra.

(i) A nonempty subset $\mathcal{F}$ of B is a *filter* if

    (a) $\mathcal{F}$ is *upward closed*, that is, $(\forall a \in \mathcal{F})(\forall b \in \mathcal{B})[a \leq b \rightarrow b \in \mathcal{F}]$, and

    (b) *closed under meet*, that is, $(\forall a \in \mathcal{F})(\forall b \in \mathcal{F})[a \wedge b \in \mathcal{F}]$,

    (c) $0 \notin \mathcal{F}$.

(ii) A nonempty $\mathcal{I} \subseteq \mathcal{B}$ is an *ideal* if it is downward closed, closed under join, and does not contain 1.

(iii) For a set U, a filter or ideal on $\mathcal{P}(U)$ is said to be filter or ideal on the set U.

It should be clear that the notions of filter and ideal are in a sense *dual*: if $\mathcal{F}$ is a filter on a Boolean algebra $\mathcal{B}$, then $\mathcal{I} = \{-b : b \in \mathcal{F}\}$ is an ideal on $\mathcal{B}$, and moreover, if $\mathcal{I}$ is an ideal on $\mathcal{B}$, then $\mathcal{F} = \{-b : b \in \mathcal{I}\}$ is a filter on $\mathcal{B}$. A filter typically serves as a measure of largeness of a subset of x, while an ideal serves as a notion of smallness.

Example 5.3.2. The *Fréchet ideal* on an infinite set U is the collection of all finite subsets of U. Similarly, the *Fréchet filter* on an infinite set U is the collection of all cofinite subsets of U.

Example 5.3.3. The *density zero ideal* on ω is the set of all sets $A \subseteq \omega$ whose upper asymptotic density $\limsup_n \frac{\text{card}(A \cap n)}{n}$ is equal to zero.

Definition 5.3.4. For a theory Γ, let $\mathcal{F}_\Gamma = \{[P] : P \in DC(\Gamma)\}$, where $DC(\Gamma)$ is the deductive closure of Γ.

Proposition 5.3.5. *Let Γ be a set of sentences in some propositional or first-order language. Then Γ is a consistent theory if and only if the deductive closure $\mathcal{F}_\Gamma$ is a filter in the corresponding Lindenbaum algebra.*

Proof. Assume first that Γ is a consistent theory. For (1), suppose that $[P] \in \mathcal{F}_\Gamma$ and $[P] \leq [Q]$. Then $P \in \Gamma$ and $P \vdash Q$, by Proposition 5.2.4. Since Γ is a theory, $Q \in DC(\Gamma)$ and hence $[Q] \in \mathcal{F}_\Gamma$. For (2), suppose that $[P] \in \mathcal{F}_\Gamma$ and $[Q] \in \mathcal{F}_\Gamma$.

Then $P \in DC(\Gamma)$ and $Q \in DC(\Gamma)$, and hence $P \wedge Q \in DC(\Gamma)$. Thus $[P] \wedge [Q] = [P \wedge Q] \in \mathcal{F}_\Gamma$. For (3), $0 = [\bot] \notin \mathcal{F}_\Gamma$ since Γ is consistent.

Next assume that $\mathcal{F}_\Gamma$ is a filter and suppose $\Gamma \vdash Q$. Then there is a finite set $\{P_1, \ldots, P_n\} \subseteq \Gamma$ such that $\{P_1, \ldots, P_n\} \vdash Q$ and hence $P_1 \wedge \cdots \wedge P_n \vdash Q$. Now each $[P_i] \in \mathcal{F}_\Gamma$ by definition and it follows from closure under meet that $[P_1 \wedge \cdots \wedge P_n] = [P_1] \wedge \cdots \wedge [P_n] \in \mathcal{F}_\Gamma$. Since $P_1 \wedge \cdots \wedge P_n \vdash Q$, it follows that $[P_1 \wedge \cdots \wedge P_n] \leq [Q]$. Thus $[Q] \in \mathcal{F}_\Gamma$ by upward closure. This now implies that $Q \in \Gamma$. $\qquad \square$

The notions of a maximal ideal and a maximal filter are very important.

Definition 5.3.6. A filter $\mathcal{M}$ on a Boolean algebra $\mathcal{B}$ is a *maximal filter* (or an *ultrafilter*) if there is no filter $\mathcal{F}$ with $\mathcal{M} \subsetneq \mathcal{F} \subsetneq \mathcal{B}$. The ideal dual to an ultrafilter is a *maximal ideal*.

Proposition 5.3.7. *A filter $\mathcal{F}$ on a Boolean algebra $\mathcal{B}$ is maximal if and only if, for every $b \in \mathcal{B}$, $b \in \mathcal{F}$ or $-b \in \mathcal{F}$. Similarly, an ideal $\mathcal{I}$ is maximal if and only if, for every $b \in \mathcal{B}$, $b \in \mathcal{I}$ or $-b \in \mathcal{I}$.*

Proof. Suppose first that $\mathcal{F}$ is a filter such that, for every $b \in \mathcal{B}$, $b \in \mathcal{F}$ or $-b \in \mathcal{F}$. If $\mathcal{F} \subsetneq \mathcal{G}$, let $b \in \mathcal{G} \setminus \mathcal{F}$. Since $b \notin \mathcal{F}$, it follows that $-b \in \mathcal{F}$ and therefore $0 = -b \wedge b \in \mathcal{G}$, a contradiction. Next suppose that $\mathcal{F}$ is maximal and that $a \notin \mathcal{F}$ and $-a \notin \mathcal{F}$. Define $\mathcal{G} = \{x \in \mathcal{B} : a \wedge b \leq x \text{ for some } b \in \mathcal{F}\}$. We claim that $\mathcal{G}$ is a proper extension of $\mathcal{F}$. Since $\mathcal{F}$ is nonempty we have some $b \in \mathcal{F}$ and $a \wedge b \leq a$, so that $a \in \mathcal{G}$. Thus $\mathcal{G}$ is a proper extension. Next we verify that $\mathcal{G}$ is a filter. It is easy to see that $\mathcal{G}$ is closed upwards and closed under meet. Suppose now that $0 \in \mathcal{G}$. Then there is $b \in \mathcal{F}$ such that $a \wedge b = 0$ and therefore $b \leq -a$, which implies that $-a \in \mathcal{F}$, a contradiction. It follows that $\mathcal{G}$ is a filter, which contradicts the maximality of $\mathcal{F}$. Thus we must have either $a \in \mathcal{F}$ or $-a \in \mathcal{F}$.

The proof for a maximal ideal is similar. $\qquad \square$

Here is another equivalent definition.

Definition 5.3.8. An ideal $\mathcal{I}$ on a Boolean algebra $\mathcal{B}$ is a *prime ideal* if for all $a, b \in \mathcal{B}$, if $a \wedge b \in \mathcal{I}$, then either $a \in \mathcal{I}$ or $b \in \mathcal{I}$.

Proposition 5.3.9. *An ideal $\mathcal{I}$ on a Boolean algebra is a* prime *ideal if and only if it is maximal.*

Proof. Suppose first that $\mathcal{I}$ is prime. Since $\mathcal{I}$ is nonempty and downward closed, $0 \in \mathcal{I}$. Then for any $a \in \mathcal{B}$, $a \wedge -a = 0 \in \mathcal{I}$ and therefore either $a \in \mathcal{I}$ or $-a \in \mathcal{I}$.

Suppose next that $\mathcal{I}$ is maximal and that $a \wedge b \in \mathcal{I}$. We will now suppose, by way of contradiction, that $a \notin \mathcal{I}$ and $b \notin \mathcal{I}$. It follows that $-a \in \mathcal{I}$ and $-b \in \mathcal{I}$. Then $-a \vee -b \in \mathcal{I}$ since $\mathcal{I}$ is an ideal. But $-a \vee -b = -(a \wedge b)$, so that the join $(a \wedge b) \vee -(a \wedge b) = 1 \in \mathcal{I}$, which is a contradiction. $\qquad\square$

How do we find an ultrafilter? There are some trivial ultrafilters.

Definition 5.3.10. For any element a of a Boolean algebra $\mathcal{B}$, let $\mathcal{F}_a = \{b : a \leq b\}$. It is easy to see that this is a filter — see the exercises. A filter $\mathcal{F}$ is *principal* if $\mathcal{F} = \mathcal{F}_a$ for some a. Similarly an ideal $\mathcal{I}$ is principal if $\mathcal{I} = \{b : b \leq a\}$ for some a.

Proposition 5.3.11. *The filter $\mathcal{F}_a$ is an ultrafilter if and only if a is an atom.*

Proof. This is left as an exercise. $\qquad\square$

For example, in $\mathcal{P}(U)$, $\mathcal{F}_a = \{b : a \subseteq b\}$ and this is an ultrafilter if and only if $a = \{x\}$ for some $x \in a$. For any finite Boolean algebra, every filter is principal. This is left as an exercise.

For most purposes, principal ultrafilters are not very interesting. To obtain nonprincipal ultrafilters, we need to use the Axiom of Choice, in the following equivalent form.

Lemma 5.3.12 (Kuratowski–Zorn Lemma). *Let P be a family of sets such that for any linearly ordered subset $\{S_i : i \in I\}$ of P, $\bigcup_{i \in I} S_i \in P$. Then P contains a $\subseteq$-maximal set M.*

Theorem 5.3.13 (Prime Ideal Theorem).

(1) *Any filter $\mathcal{F}_0$ on a Boolean algebra may be extended to a maximal filter.*
(2) *Any ideal $\mathcal{I}_0$ on a Boolean algebra may be extended to a maximal ideal.*

Proof. Let P be the family of filters $\mathcal{F}$ such that $\mathcal{F}_0 \subseteq \mathcal{F}$, ordered by inclusion ($\subseteq$). Let $S = \{\mathcal{F}_i : i \in I\}$ be any linearly ordered subset of P and let $\mathcal{G} = \bigcup_{i \in I} \mathcal{F}_i$. To apply Kuratowski's lemma, we need to check that $\mathcal{G} \in P$, that is, $\mathcal{G}$ is a filter which extends $\mathcal{F}$.

$\mathcal{F}_0 \subseteq \mathcal{G}$: For each $i \in I$, $\mathcal{F}_0 \subseteq \mathcal{F}_i$, so that $\mathcal{F}_0 \subseteq \bigcup_{i \in I} \mathcal{F}_i$.

$\mathcal{G}$ is upward closed: Suppose $a \in \mathcal{G}$ and $a \leq b$. Then $a \in \mathcal{F}_i$ for some $i \in I$. Thus $b \in \mathcal{F}_i$, since $\mathcal{F}_i$ is a filter. It follows that $b \in \bigcup_{i \in I} \mathcal{F}_i$.

$\mathcal{G}$ is closed under meet: Suppose $a \in \mathcal{G}$ and $b \in \mathcal{G}$. Then for some i and j, $a \in \mathcal{F}_i$ and $b \in \mathcal{F}_j$. Since S is linearly ordered, we may assume without loss of generality that $\mathcal{F}_i \subseteq \mathcal{F}_j$. Then $a, b \in \mathcal{F}_j$, so that $a \wedge b \in \mathcal{F}_j$, since $\mathcal{F}_j$ is a filter. Thus $a \wedge b \in \mathcal{G}$.

$0 \notin \mathcal{G}$: For each i, $0 \notin \mathcal{F}_i$ and therefore $0 \notin \bigcup_{i \in I} \mathcal{F}_i$.

Now by Kuratowski's lemma, P contains a maximal element $\mathcal{M}$. Then $\mathcal{M}$ is a filter extending $\mathcal{F}$ which has no proper extension in P; that is, $\mathcal{M}$ is a maximal filter.

The proof for ideals is similar, or one can use the duality of ideals mentioned above — this is left as an exercise. $\square$

We are most interested in ideals and filters on the Boolean algebra $\mathcal{P}(U)$ for a given infinite set U. Recall from above that an atom a in $\mathcal{P}(U)$ is just a singleton $a = \{u\}$ for some $u \in U$.

Theorem 5.3.14. *There is a nonprincipal ultrafilter on every infinite set.*

Proof. Let an infinite set X be given and let $\mathcal{F}$ be the Fréchet filter of cofinite subsets of X. An application of the Prime Ideal Theorem (Theorem 5.3.13) gives us an ultrafilter $\mathcal{U}$ on $\mathcal{P}(X)$ extending $\mathcal{F}$. If $\mathcal{U}$ were principal, then by Proposition 5.3.11, $\mathcal{U}$ would equal $\mathcal{F}_{\{x\}}$ for some $x \in X$. This would mean that the finite set $\{x\}$ belongs to $\mathcal{U}$. But $X \setminus \{x\}$ belongs to $\mathcal{U}$ since $\mathcal{U}$ extends the Fréchet filter, and then $\emptyset = \{x\} \cap (X \setminus \{x\})$, so that $\mathcal{U}$ is not a filter. $\qquad\square$

For more about the Axiom of Choice, see Chapter 6 of the companion volume on set theory [4].

Exercises for Section 5.3

Exercise 5.3.1. Show that the condition of being upward closed may be replaced in the definition of a filter by the following: If $a \in \mathcal{F}$ and $b \in \mathcal{B}$, then $a \vee b \in \mathcal{F}$.

Exercise 5.3.2. Show that if $\mathcal{F}$ is a filter on $\mathcal{B}$, then $\{b : -b \in \mathcal{F}\}$ is an ideal.

Exercise 5.3.3. Use the duality of ideals and filters to show that any ideal $\mathcal{I}$ on a Boolean algebra may be extended to a maximal ideal, assuming the corresponding result for filters.

Exercise 5.3.4. Show that if an ultrafilter on $\mathcal{P}(U)$ contains a finite set, then it is in fact principal.

Exercise 5.3.5. Show that a filter $\mathcal{F}$ on a Boolean algebra $\mathcal{B}$ is an ultrafilter if and only if, for any two elements a and b, if $a \vee b \in F$, then either $a \in F$ or $b \in F$.

Exercise 5.3.6. Show that for a finite Boolean algebra $\mathcal{B}$, every filter on $\mathcal{B}$ has the form $\mathcal{F}_a$ for some a.

Exercise 5.3.7. Show that the filter $\mathcal{F}_a$ is an ultrafilter if and only if a is an atom.

Exercise 5.3.8. Show that for any nonzero element a of a Boolean algebra $\mathcal{B}$, $\mathcal{F}_a = \{b : a \leq b\}$ is a filter.

5.4.   Ultraproducts

In this section, we introduce the notion of an ultraproduct of a sequence of structures and explore the properties of ultraproducts. Ultraproducts will be used later to construct nonstandard models of the natural numbers and of the real numbers. The existence of ultrafilters is applied to obtain ultraproducts. For a thorough development of ultraproducts, see [6] or [1].

Definition 5.4.1. Let U be an ultrafilter on a set I; I is called an *index* set. Let A be any set and define a relation $\sim_U$ (*equality mod U*) on the family A^I of functions from I to A, by

$$x \sim_U y \iff \{i \in I : x(i) = y(i)\} \in U,$$

and let $[x]_U$ be the equivalence class of x under this relation.

We say that $x(i)$ equals $y(i)$ *almost always* when $x \sim_U y$.

Proposition 5.4.2. *Given an ultrafilter U on a set I and a set A, the relation $\sim_U$, equality mod U, is an equivalence relation on A^I.*

Proof.   The proof consists of the following three claims.

(a) The relation $\sim$ is reflexive. That is, $x \sim x$ for all x.
(b) The relation $\sim$ is symmetric. That is, if $x \sim y$, then $y \sim x$.
(c) The relation $\sim$ is transitive. That is, if $x \sim y$ and $y \sim z$, then $x \sim z$.

Here is the proof of part (a). For any $x \in A^I$, $\{i \in I : x(i) = x(i)\} = I$ and $I \in U$ for any filter U. The other cases are left as exercises.   □

Definition 5.4.3. Let $\mathcal{L}$ be first-order language. Suppose that $\mathcal{A}_i$ is an $\mathcal{L}$-structure with domain A_i for each $i \in I$ and that U

is an ultrafilter on I. Let $[f]_U$ denote the equivalence class of $f \in \bigotimes_{i \in I} A_i$ under the equivalence relation $\sim_U$ (where $\bigotimes_{i \in I} A_i$ is the direct product of the sets $(A_i)_{i \in I}$). Then the *ultraproduct* $\mathcal{C} = \bigotimes_{i \in I} A_i / U$ is defined as follows:

(a) $\mathcal{C}$ has universe $C = \{[f]_U : f \in \bigotimes_{i \in I} A_i\}$.
(b) For each constant symbol c of $\mathcal{L}$, $c^{\mathcal{C}} = [f]_U$, where $f(i) = c^{A_i}$ for each $i \in I$.
(c) For each n-ary function symbol $F \in \mathcal{L}$, $F^{\mathcal{C}}([f_1]_U, \ldots, [f_n]_U) = [g]_U$, where $g(i) = F^{A_i}(f_1(i), \ldots, f_n(i))$.
(d) For each n-ary relation symbol $R \in \mathcal{L}$,

$$R^{\mathcal{C}}([f_1]_U, \ldots, [f_n]_U) \iff \{i \in I : R^{A_i}(f_1(i), \ldots, f_n(i))\} \in U.$$

If there is a fixed structure $\mathcal{A}$ such that $\mathcal{A}_i = \mathcal{A}$ for all $i \in I$, then $\bigotimes_{i \in I} A_i / U$ is also written as $\mathcal{A}^I / U$ and is an *ultrapower*.

Example 5.4.4. Let U be the principal ultrafilter generated by $\{1\}$ in $\mathcal{P}(3)$. Then $(\{0, 1\} \times \{0, 1\} \times \{5\})/U$ has exactly two elements as a structure over the language with no predicate letters, function symbols or constants.

Notice that $\{0, 1\} \times \{0, 1\} \times \{5\} = \{(0, 0, 5), (0, 1, 5), (1, 0, 5), (1, 1, 5)\}$. Since $\{1\}$ is in U, $(0, 0, 5) \sim (1, 0, 5)$ and $(0, 1, 5) \sim (1, 1, 5)$. Thus the $(\{0, 1\} \times \{0, 1\} \times \{5\})/U = \{[(0, 0, 5)], (0, 1, 5)]\}$.

Example 5.4.5. If U is a nonprincipal ultrafilter on ω, then the ultraproduct $\bigotimes_{i \in \omega} \{0, 1\}/U$ has exactly two elements, even though the index set is infinite.

For each natural number i, let $\mathcal{A}^i = \{0, 1\}$. The set $\bigotimes_{i \in \omega} \mathcal{A}$ has uncountably many elements, one for each real number. For $e \in \{0, 1\}$, let $g_e \in \{0, 1\}^\omega$ be the constant function with $g_e(i) = e$ for all $i \in \omega$. Given any function g in $\bigotimes_{i \in \omega} \mathcal{A}$, consider $\text{Zero}(g) = \{i \in \omega \mid g(i) = 0\}$. Then either $\text{Zero}(g)$ is in U or $\omega \setminus \text{Zero}(g)$ is in U. If $\text{Zero}(g)$ is in U, then g is equivalent to g_0. Otherwise, g is equivalent to g_1. So the product has only these two equivalence classes.

Theorem 5.4.6 (Łoś's Theorem). *Let $\mathcal{L}$ be a first-order language and let $\{\mathcal{A}_i : i \in I\}$ be an indexed family of $\mathcal{L}$-structures for some index set I. Let U be an ultrafilter on index set I. Then for any formula φ,*

$$\bigotimes_{i \in I} \mathcal{A}_i/U \models \varphi(f_1, \ldots, f_n)$$

$$\iff \{i \in I : \mathcal{A}_i \models \varphi(f_1(i), \ldots, f_n(i))\} \in U.$$

Proof. The proof proceeds by induction on formulas. The first three claims follow from the definition of an ultraproduct. Let $\mathcal{B} = \bigotimes_{i \in I} \mathcal{A}_i/U$.

(a) For all terms t, $t^{\mathcal{B}} = [f_t]_U$, where $f_t(i) = t^{\mathcal{A}_i}$.

(b) For all terms s and t, $\mathcal{B} \models s = t$ if and only if $\{i \in I : \mathcal{A}_i \models s = t\} \in U$.

(c) For all predicate letters R and all terms $t_1, \ldots, t_r$, $\mathcal{B} \models R(t_1, \ldots, t_r)$ if and only if $\{i \in I : \mathcal{A}_i \models P(t_1, \ldots, t_r)\} \in U$.
Let BOTH be the set of formulas θ for which $\mathcal{B} \models \theta$ if and only if $\{i \in I : \mathcal{A}_i \models \theta\} \in U$.

(d) BOTH is closed under the connectives $\neg$, $\wedge$.

(e) BOTH is closed under forming new formulas via the existential quantifier. For simplicity, let θ be the formula $(\exists x)\psi(x)$ with parameters suppressed. Suppose first that $J = \{i : \mathcal{A}_i \models (\exists x)\psi(x)\} \in U$. Define f so that $\mathcal{A}_i \models \psi(f(i))$ whenever $i \in J$ and let $f(i) \in \mathcal{A}_i$ be arbitrary otherwise. Then $\{i : \mathcal{A}_i \models \psi(f(i)\} = J$ and belongs to U. It follows by induction that $\mathcal{B} \models \psi([f]_u)$ and therefore $\mathcal{B} \models (\exists x)\psi(x)$.
Next suppose that $\mathcal{B} \models (\exists x)\psi(x)$ and choose f so that $\mathcal{B} \models \psi([f]_U)$. It follows by induction that $J = \{i : \mathcal{A}_i \models \psi(f(i))\} \in U$. It further follows by induction that, for each $i \in J$, $\mathcal{A}_i \models (\exists x)\psi(x)$ and therefore $J \subseteq \{i : \mathcal{A} \models (\exists x)\psi(x)\}$, so the latter set is also in U.

The proof of part (d) is left as an exercise. $\qquad\square$

An important consequence of Theorem 5.4.6 is that for any fixed structure $\mathcal{A}$, $\mathcal{A}$ may be viewed as an elementary submodel of the ultrapower $\bigotimes_{i \in I} \mathcal{A}/U$. Let a structure $\mathcal{A}$ with universe A be given. Then, for each element $a \in A$, let $c_a : I \to A$ be the constant function defined by $c_a(i) = a$ for all $i \in I$.

Corollary 5.4.7. *Let $\mathcal{A}$ be any structure, let U be an ultrafilter on the set I, and let $\mathcal{B}$ be the ultrapower $\bigotimes_{i \in I} \mathcal{A}/U$. Then for any formula φ and any elements $a_1, \ldots, a_n \in \mathcal{A}$,*

$$\mathcal{A} \models \varphi(a_1, \ldots, a_m) \iff \mathcal{B} \models \varphi([c_{a_1}]_U, \ldots, [c_{a_n}]_U),$$

where we identify $a \in \mathcal{A}$ with the equivalence class $[c_a] \in \mathcal{B}$ of the constant function c_a.

Proof. Let us denote $[c_a]$ by $[a]$. By Theorem 5.4.6, $\mathcal{A}^* \models \varphi([a_1], \ldots, [a_n])$ if and only if $\{i \in I \mid \mathcal{A} \models \varphi(a_1(i), \ldots, a_n(i))\} \in U$, which holds if and only if $\mathcal{A} \models \varphi(a_1, \ldots, a_n)$, since the given set is either I if $\mathcal{A} \models \varphi(a_1, \ldots, a_n)$ or is $\emptyset$ if $\mathcal{A} \models \neg\varphi(a_1, \ldots, a_n)$. $\square$

Proposition 5.4.8. *If $|A_i| < |A_{i+1}|$ for all i in ω, and U is nonprincipal, then $\bigotimes_{i \in \omega} A_i/U$ is infinite.*

Proof. If $|A_i| < |A_{i+1}|$ for all i in ω, then by induction on i, one sees that for all i in ω, $|A_i| \geq i$. Therefore for each i in ω, there is a one-to-one mapping f_i from $\{1, 2, \ldots, i\}$ to A_i. For each n in ω, define a function g_n as follows:

$$g_n(m) = \begin{cases} f_m(n) & \text{if } n \leq m, \\ f_m(1) & \text{otherwise.} \end{cases}$$

Here is a pictorial representation of the functions g_n.

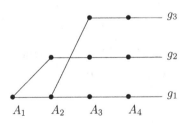

Notice that for $m \geq n, \ell$, if $n \neq \ell$, then since $g_n(m) = f_m(n)$, $g_\ell(m) = f_m(\ell)$ and f_m is one-to-one, it follows that $g_n(m) \neq g_\ell(m)$. Thus there are at most finitely many m on which $g_n(m)$ and $g_\ell(m)$ agree. Therefore, since U is nonprincipal, the equivalence classes are distinct, $[g_n]_U \neq [g_\ell]_U$. Consequently, $\bigotimes_{i \in \omega} \mathcal{A}_i / U$ is infinite. □

Here is another corollary of Łoś's Theorem.

Corollary 5.4.9. *If φ is a sentence with arbitrarily large finite models, then φ has an infinite model.*

Proof.   Suppose that the sentence φ has arbitrarily large finite models and let $\mathcal{A}_i$ be a model of φ with at least i elements. Let U be a nonprincipal ultrafilter on ω and let $\mathcal{A} = \bigotimes_{i \in \omega} \mathcal{A}_i / U$. Since each $\mathcal{A}_i \models \varphi$, it follows from Łoś's theorem that $\mathcal{A} \models \varphi$. Recall the sentence $\mathcal{E}^{\geq n}$ such that a structure $\mathcal{B} \models \mathcal{E}^{\geq n}$ if and only if $\mathcal{B}$ has at least n elements. Then for each n, $\{i : \mathcal{A}_i \models \mathcal{E}^{\geq n}\}$ includes the cofinite set $\{i : i \geq n\}$ and therefore belongs to U. It follows again from from Łoś's theorem that $\mathcal{A} \models \mathcal{E}^{\geq n}$ for each n and hence $\mathcal{A}$ is an infinite model of φ. □

Note that we have looked at this problem using the Compactness Theorem. See Example 4.17.

Corollary 5.4.10. *There is no first-order sentence φ so that for every structure $\mathcal{A}$, $\mathcal{A}$ is finite if and only if $\mathcal{A} \models \varphi$.*

The proof is left as an exercise.

Exercises for Section 5.4

Exercise 5.4.1. Show that if $\mathcal{A}$ is finite, then for any ultrafilter U, $\bigotimes_{i \in I} \mathcal{A}/U$ is isomorphic to $\mathcal{A}$.

Exercise 5.4.2. Show that if U is a principal ultrafilter on $\mathbb{N}$, then for any family of structures $\{\mathcal{A}_i : i \in \mathbb{N}\}$, $\bigotimes_{i \in \mathbb{N}} \mathcal{A}_i / U$ is isomorphic to $\mathcal{A}_i$ for some $i \in I$.

Exercise 5.4.3. Complete the proof of Proposition 5.4.2 by showing that the relation $\sim_U$ is reflexive and transitive.

Exercise 5.4.4. Show that the set BOTH defined in the proof of Łoś's theorem is closed under the connectives $\neg$, $\wedge$.

Exercise 5.4.5. For each prime p, the finite field of integers with arithmetic mod p is a field,

$$\mathbb{Z}_p = \langle \{\, 0, 1, \ldots, p-1 \,\}, +_{\mathrm{mod}\ p}, \times_{\mathrm{mod}\ p}, 0, 1 \rangle.$$

We say that $\mathbb{Z}_p$ has *characteristic* p since it satisfied the sentence

$$\mathbb{Z}_p \models (\forall x)(\underbrace{x + x + \cdots + x}_{p \text{ times}} = 0).$$

If a field does not have finite characteristic, it is said to have *characteristic* 0. Suppose that $\{p(n) \colon n \in \omega\}$ is a list of all prime numbers in increasing order and U is a nonprincipal ultrafilter on ω.

(1) What is the characteristic of $\bigotimes_{n \in \omega} \mathbb{Z}_{p(n)}/U$?
(2) What is the characteristic of $\bigotimes_{n \in \omega} \mathbb{Z}_2/U$?

Exercise 5.4.6. Show that there is no first-order sentence φ so that for every structure $\mathcal{A}$, $\mathcal{A}$ is finite if and only if $\mathcal{A} \models \varphi$.

Chapter 6

Computability

The theory of computability traces back to the work of Gödel, Church, Tulring and others in the 1930s. There are many different approaches to defining the collection of computable number-theoretic functions. The intuitive idea is that a function $F : \mathbb{N} \to \mathbb{N}$ (or more generally $F : \mathbb{N}^k \to \mathbb{N}$) is computable if there is an algorithm or effective procedure for determining the output $F(m)$ from the input m. To demonstrate that a particular function F is computable, it suffices to give the corresponding algorithm. We will begin by considering examples of intuitively computable functions; we will later consider formal definitions of computable functions (and it will follow that the functions given below are all computable in this formal sense).

Example 6.0.1. Basic computable functions include

  (i) the successor function $S(x) = x + 1$,
  (ii) the addition function $+(x, y) = x + y$, and
 (iii) the multiplication function $\cdot \, (x, y) = x \cdot y$.

Example 6.0.2. Here are some slightly more complicated examples of computable functions:

  (i) The Division Algorithm demonstrates that the two functions that compute, for inputs a and b, the unique quotient

$q = q(a, b)$ and remainder $r = r(a, b)$, with $0 \le r < a$, such that $a = qb + r$, are both computable.

(ii) The Euclidean Algorithm demonstrates that the function $\gcd(a, b)$, which computes the greatest common divisor of a and b, is computable. It follows that least common multiple function $\text{lcm}(a, b) = (a \cdot b)/\gcd(a, b)$ is also computable

The notion of computability for functions can be extended to subsets of $\mathbb{N}$ or relations on $\mathbb{N}^k$ for some k as follows. First, a set $A \subseteq \mathbb{N}$ is said to be computable if the characteristic function of A, defined by

$$\chi_A(n) = \begin{cases} 1 & \text{if } n \in A, \\ 0 & \text{if } n \notin A, \end{cases}$$

is a computable function. Similarly, a relation $R \subseteq \mathbb{N}^k$ is said to be computable if its characteristic function

$$\chi_A(n_1, \ldots, n_k) = \begin{cases} 1 & \text{if } (n_1, \ldots, n_k) \in R \\ 0 & \text{if } (n_1, \ldots, n_k) \notin R \end{cases}$$

is computable. These definitions are equivalent to saying that there is an algorithm for testing whether a given number is in A or whether a given finite sequence $(n_1, \ldots, n_k)$ is in R.

Example 6.0.3. The following conditions are computable:

(i) The set of perfect squares is computable, since given a number n, we can test whether it is a square by computing m^2 for all $m \le n$ and checking whether $m^2 = n$.

(ii) The relation $x \mid y$ (x divides y) is computable, since by the Division Algorithm, $x \mid y$ if and only if the remainder $r(x, y) = 0$.

(iii) The set of even numbers is computable, since n is even if and only if $2 \mid n$.

(iv) The set of prime numbers is computable, since p is prime if and only if

$$(\forall m < p)[(m \ne 0 \wedge m \ne 1) \rightarrow m \nmid p].$$

Two formalizations of the class of computable functions that will we consider are the collection of Turing machines and the collection of partial recursive functions. All of the formalizations of the intuitive notion of a computable number-theoretic function has given rise to the same collection of functions. For this and other reasons, the community of mathematical logicians has largely come to accept the *Church–Turing thesis*, which is the claim that the collection of Turing computable functions is the same as the collection of intuitively computable functions. In practice, the Church–Turing thesis has two main consequences:

(1) if we want to show that a given problem cannot be solved by any algorithmic procedure, it suffices to show that solutions to the problem cannot be computed by any Turing computable functions;
(2) to show that a given function is computable, it suffices to give an informal description of an effective procedure for computing the values of the function.

We will see that it is important to extend the notion of computability to functions on *strings* (or *words*) over a finite alphabet Σ.

Definition 6.0.4.

(1) For a fixed alphabet Σ, a *language* is simply a set $L \subseteq \Sigma^*$.
(2) The *length* $|w|$ of a string $w = w_0 w_1 \cdots w_{n-1}$ is n. The *empty word* ε has length zero.
(3) For two strings $w_1 = a_0 a_1 \cdots a_{m-1}$ and $w_2 = b_0 b_1 \cdots b_{n-1}$, the *concatenation* $w_1 \frown w_2$ (sometimes just written $w_1 w_2$) is the string

$$w_1 \frown w_2 = a_0 a_1 \cdots a_{m-1} b_0 b_1 \cdots b_{n-1}.$$

Here we are using the terminology of computer science, so that a language may be any set of words.

Example 6.0.5. Recall that for the propositional calculus, the language $\mathcal{L}(\{A_1, \ldots, A_n\})$ was the set of propositional sentences built up from the propositional variables $A_1, \ldots, A_n$ as well as the symbols $\{\neg, \vee, \wedge, \rightarrow\}$. This is also a language in the sense given above and is in fact computable. However, the language $\{A_1, \ldots, A_n, \neg, \vee, \wedge, \rightarrow\}^*$ also includes strings such as $A_1 \neg \wedge$ which are not sentences.

Example 6.0.6. Let Σ be any finite alphabet and let $a \in \Sigma$.

(i) Define the function S_a by $S_a(w) = wa$. Then S_a is computable.
(ii) The concatenation function defined above is computable.

Here are some computable sets of words.

Example 6.0.7.

(i) The set of tautologies of propositional logic over two propositional variables $\{A, B\}$ is computable.
(ii) The set of words $w \in \{0, 1\}^*$ such that every 0 in w is followed by a 1.

For a thorough study of computability, see books of Cooper [7], Sipser [29], and Soare [30].

6.1.   Finite State Machines

As a warm-up, we will first consider a simplified version of a Turing machine, known as a finite state machine, also known as a finite state automaton.

Definition 6.1.1. A *finite state machine* (*FSM*) over a finite alphabet Σ (usually $\{0, 1\}$) is given by the following conditions:

(i) a finite set Q of states;
(ii) a transition function $\delta : Q \times \Sigma \rightarrow Q$;
(iii) an *initial state* in Q and a set $A \subseteq Q$ of *accepting* states.

Definition 6.1.2. The computation of an FSM M on input $w = a_0 a_1 \cdots a_k$ is a sequence of stages of length k.

- At stage 0, the machine begins in the initial state s_0, scans the input a_0, and then and transitions to state $s_1 = \delta(s_0, a_0)$.
- At stage i, the machine (in state s_i) scans a_i and transitions to state $s_{i+1} = F(s_i, a_i)$.
- After reading a_k during stage k, the machine halts in state s_{k+1}. The input word w is *accepted* by M if $s_{k+1} \in A$.
- The language $L(M)$ is the set of words accepted by M.

Definition 6.1.3. A subset L of Σ^* is said to be a *regular language* if $L = L(M)$ for some FSM M.

Example 6.1.4. Let M_1 be the finite state machine, with alphabet $\{0, 1\}$ and transition function

$$\delta(q_i, j) = \begin{cases} q_{1-i} & \text{if } i = j, \\ q_i & \text{if } i \neq j, \end{cases}$$

for $i, j \in \{0, 1\}$. Then $L(M_1)$ is the set of words which end in a 0.

Finite state machines are often visualized by *state diagrams*. To illustrate the idea, consider the following diagram for the machine M_1.

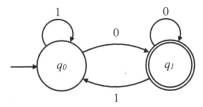

States are represented as nodes, with labeled edges representing the transition function between states. The arrow entering the state q_0 indicates that q_0 is the initial state, and the doubled circles indicate that q_1 is an accepting state.

Example 6.1.5. Let M_2 be the FSM, depicted by the state diagram below, with transition function

$$\delta(q_i, j) = \begin{cases} q_{1-i} & \text{if } j = 0, \\ q_i & \text{if } j = 1, \end{cases}$$

for $i = 0, 1$. Thus $L(M_2)$ is the set of words which contain an even number of 0's.

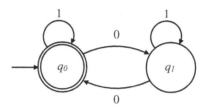

There are languages with very simple definitions which are not regular.

Proposition 6.1.6. $L = \{a^n b^n : n \in \omega\}$ *is not a regular language.*

Proof. Suppose by way of contradiction that $L = L(M)$ for some FSM M and let k be the number of states of M. Consider the sequence of states $s_0, s_1, \ldots, s_k$ resulting when M reads the input a^k. It follows from the pigeonhole principle that $s_i = s_j$ for some $i < j \le k$. Thus M ends up in the same state s_i after reading a^i and after reading a^j. But this means that M ends up in the same state q after reading $a^i b^i$ that it ends up in after reading $a^j b^i$. By assumption, M accepts $a^i b^i$, so that q is an accepting state. However, M does not accept $a^i b^j$, so that q is not accepting. This contradiction shows that $L \ne L(M)$. □

Next we consider some finite state machines which have an output function. These are called finite state *transducers*.

Definition 6.1.7. A finite state transducer (FST) M is a FSM which has in addition to (i), (ii), (iii) above,

(iv) an output function $F : Q \times \Sigma \to \Sigma^*$.

At each stage i in a computation by an FST, we can think of the machine as writing an output symbol $b_i = F(s_i, a_i)$ on a tape; this may be the input symbol which was already there. When M halts in the final state after reading the input word w, the output word $b_0 b_1 \cdots b_k$ written on the tape is the output $M(w)$.

We want to express natural numbers in reverse binary notation, so that the word $a_0 a_1 \cdots a_k$ represents $a_0 + 2a_1 + \cdots + 2^k a_k$.

Example 6.1.8. The successor machine M_3 computes $S(x) = x + 1$. State q_0 is the *carry* state and state q_1 is the *no-carry* state. The edges in the state diagram below are labeled with symbols of the form $i : j$, which means that i is an input bit and j is an output bit.

For reasons that will be clear shortly, we require that any number of the form $2^n - 1$ (normally represented by a string of the form 1^n) to be represented by $1^n 0$. For any other number, adding additional zeros to the end of its representation will make no difference.

Starting in state q_0, the machine outputs $1 - i$ on input i and transitions to state q_{1-i}. From state q_1 on input i, the machine outputs i and remains in state q_1. We take the liberty of adding an extra 0 at the end of the input in order to accommodate the carry.

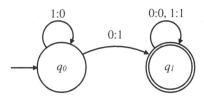

More precisely, the transition function and output function of M_3 are defined by

$$\delta(q_i, j) = \begin{cases} q_0 & \text{if } i = 0 \wedge j = 1, \\ q_1 & \text{if } (i = 0 \wedge j = 0) \vee i = 1 \end{cases}$$

and

$$F(q_i, j) = \begin{cases} 1 - j & \text{if } i = 0, \\ j & \text{if } i = 1. \end{cases}$$

For $s = a_0 a_1 \cdots a_k$ see that this computes $S(x)$ by the following reasoning. Assume that x ends with 0 as described above. We consider two cases:

- Case 1: x contains a 0 before the final 0. Let i be least such that $a_i = 0$. Then M will write i 1's and remain in state q_0 until arriving at $a_i = 0$. Then M will output $b_i = 1$ and transition to state q_1. After that it will simply write $b_j = a_j$ for all $j > i$.
- Case 2: If x consists of all 1's before the final 0, then M will write n 0's and remain in state q_0 until reaching the extra 0, when M will output a final 1 and transition to q_1.

In each case, $b_1 \cdots b_n = M(x)$ is the reverse binary representation of $S(x)$ and we will end up in an accepting state.

Using the carry and no-carry states, we can also perform addition with a finite state automaton.

Example 6.1.9. The addition machine M_4 computes $S(x, y) = x + y$. Unlike the previous example, state q_0 is the *no-carry* state and state q_1 is the *carry* state. Moreover, we work with a different input alphabet: the input alphabet consists of pairs $\binom{i}{j} \in \{0, 1\} \times \{0, 1\}$. To add two numbers n_1 and n_2 represented by σ_1 and σ_2, if σ_1 is shorter than σ_2, we append 0s to σ_1 so that the resulting string has the same length as σ_2. As in the previous example, we will also append an additional 0 to the end of both σ_1 and σ_2, which is necessary in case that the string representing $n_1 + n_2$ is strictly longer than the strings representing n_1 and n_2.

The state diagram is given by:

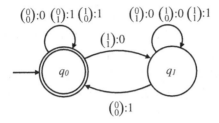

The transition function and output function of M_4 are defined in the following tables:

δ	$\begin{pmatrix} 0 \\ 0 \end{pmatrix}$	$\begin{pmatrix} 0 \\ 1 \end{pmatrix}$	$\begin{pmatrix} 1 \\ 0 \end{pmatrix}$	$\begin{pmatrix} 1 \\ 1 \end{pmatrix}$
q_0	q_0	q_0	q_0	q_1
q_1	q_0	q_1	q_1	q_1

F	$\begin{pmatrix} 0 \\ 0 \end{pmatrix}$	$\begin{pmatrix} 0 \\ 1 \end{pmatrix}$	$\begin{pmatrix} 1 \\ 0 \end{pmatrix}$	$\begin{pmatrix} 1 \\ 1 \end{pmatrix}$
q_0	0	1	1	0
q_1	1	0	0	1

Every function computed by a FST M is computable since M provides an algorithm for the computation of $M(w)$ on input w. There are functions which are clearly computable, but cannot be computed a finite state machine.

Example 6.1.10. The function $f(x) = x^2$ is not computable by a finite state transducer. Suppose that M computes x^2 on input x, where we have appended a sufficient number of 0's to the end of the input so that we can output x^2 (recall that each input bit yields at most one output bit). In particular, for any n, M will output $0^{2n}1$ on input 0^n1, since $f(2^n) = (2^n)^2 = 2^{2n}$. Thus, on input 0^n1, after reading the first $n+1$ bits, M needs to examine at least n additional 0's and write at least n additional 0's before it finishes the output with a final 1. Now suppose that M has k states and let $n > k + 1$. Let s_j be the state of the machine after reading $0^n 10^j$. Then there must be some $i, j \leq k + 1 < n$ such that $s_i = s_j$. Furthermore, every time M transitions from state s_i upon reading a 0, M must output a 0. But the machine is essentially stuck in a *loop* and hence can only print another 0 after reading $0^n 10^n$ when it needs to print the final 1.

Exercises for Section 6.1

Exercise 6.1.1. Define a finite state machine M such that $L(M)$ is the set of words from $\{0, 1\}^*$ such that all blocks of zeroes have length a multiple of three.

Exercise 6.1.2. Define a FSM M such that $L(M)$ is the set of words from $\{0, 1\}^*$ such that every occurrence of 11 is followed by 0.

Exercise 6.1.3. Show that there is no FSM M such that $L(M)$ is the set of words with an equal number of 1's and 0's.

Exercise 6.1.4. Define a FST M on the alphabet $\{0, 1, 2\}$ such that $M(w)$ is the result of erasing all 0's from w.

Exercise 6.1.5. Define a FST M such that $M(w) = 3 \cdot w$ where w is a natural number expressed in reverse binary form.

Exercise 6.1.6. Show that if L_1 and L_2 are regular languages, then $L_1 \cup L_2$ is a regular language.

Exercise 6.1.7. Show that the set $\{1^n : n \text{ is a square}\}$ is not a regular language.

Exercise 6.1.8. Show that the set of sentences of propositional logic over $\{A, B\}$ is not a regular language.

6.2. Turing Machines

The first general model of computation that we will consider is the Turing machine, developed by Alan Turing in the 1930's. A Turing machine is a simple model of a computer that is capable of computing *any* function that can be computed. It consists of these items:

(1) a finite state control component with a finite number of read/write heads; and
(2) a finite number of unbounded memory tapes (one or more for the input(s), one for the output, and the rest for scratchwork), each of which is divided into infinitely many consecutive cells in which symbols can be written.

Furthermore, it must satisfy these conditions:

(1) there is a specified initial state q_0 and a specified final q_H, or *halting*, state; and
(2) each read/write head reads one cell of each tape and either moves one cell to the left (L), moves one cell to the right (R), or stays stationary (S) at each stage of a computation.

The notion of Turing machine is formalized in the following definition.

Definition 6.2.1. A k-tape *Turing machine* M consists of

(i) a finite set Q of states;
(ii) an input alphabet Σ and a tape alphabet $\Gamma \supseteq \Sigma$ including a blank symbol $\square$ not part of the input alphabet;
(iii) a transition function $\delta : Q \times \Gamma^k \to \Gamma^k \times \{L,R,S\}^k$;
(iv) an initial state in Q; and
(v) a halting or final state in Q.

The blank symbol indicates that nothing has been written on a square. A *move* of M in a given state q_i, scanning the symbols $a_1, \ldots, a_k$ on the k tapes, where $\delta(q_i, a_1, \ldots, a_k) = (q_j, b_1, \ldots, b_k, D_1, \ldots, D_k)$, consists of the following actions:

(1) switching from state q_i to state q_j;
(2) writing b_i (and thus erasing a_i) on tape i; and
(3) moving the head on tape i in the direction D_i.

A *computation* always begins with

   (i) the machine in state q_0;
  (ii) some finite input on each of the input tapes; and
 (iii) each of the input heads scanning the first symbol on each of the input tapes.

The *configuration* of a machine at a given stage of a computation consists of

   (i) the current state of the machine;
  (ii) the contents of each tape; and
 (iii) the location of each of the heads on each tape.

M *halts* after n moves if it transitions to the halting state in the nth stage of the computation. A machine M *accepts* a word w, denoted $M(w)\!\downarrow$, if the machine halts (i.e., ends up in the halting state) when given the input w. In this case, we say M *halts* or *converges* on the input w, denoted $M(w)\!\downarrow$; we say that the computation is *convergent*. Otherwise, we say that M *diverges* on input w, denoted $M(w)\!\uparrow$; we say that the computation is *divergent*.

It is an essential feature of Turing machines that they may fail to halt on some inputs. This gives rise to the *partial computable function* f_M which has domain $\{w : M(w)\!\downarrow\}$. That is, $f_M(w) = y$ if and only if M halts on input w and y appears on the output tape when M halts, meaning that the output tape contains y surrounded by blanks on both sides. (For the sake of elegance, we may insist that the first symbol of y is scanned at the moment of halting). Sometimes we will write the value $f_M(w)$ as $M(w)$.

Example 6.2.2. We define a Turing machine M_1 that computes $x + y$. There are two input tapes and one output tape. The numbers x and y are written on separate input tapes in reverse binary form. M has the states: the initial state q_0, a *carry* state q_1 and the halting state q_H. The two input tapes are read simultaneously in the form a/b. We need to use an extra symbol # in case one input is longer and/or there is a carry at the end. That is, #'s are appended to the end of the shorter string so that the input on each tape has the same length. The behavior of M_1 is summed up in the following table.

State	Read	Write	Move	New State
q_0	0/0	0	R	q_0
q_0	0/#	0	R	q_0
q_0	#/0	0	R	q_0
q_0	0/1	1	R	q_0
q_0	1/0	1	R	q_0
q_0	1/#	1	R	q_0
q_0	#/1	1	R	q_0
q_0	1/1	0	R	q_1
q_0	#/#	#	S	q_H
q_1	0/0	1	R	q_0
q_1	0/#	1	R	q_0
q_1	#/0	1	R	q_0
q_1	0/1	0	R	q_1
q_1	1/0	0	R	q_1
q_1	1/#	0	R	q_1
q_1	#/1	0	R	q_1
q_1	1/1	1	R	q_1
q_1	#/#	1	S	q_H

Example 6.2.3. We roughly describe a Turing machine M_2 that computes $x \cdot y$. Again there are two input tapes and one output tape. The idea is that if $y = \Sigma_{i \in I} 2^i$ for some finite $I \subseteq \mathbb{N}$, then $x \cdot y = \Sigma_{i \in I} 2^i \cdot x$. For each i, $2^i \cdot x$ has the form of i 0's followed by x. For example, if $x = 1011$ (thirteen), then $4 \cdot x = 001011$ (fifty-two).

To multiply 1011 by 100101 we add

$$2^0 \cdot 13 + 2^3 \cdot 13 + 2^5 \cdot 13 = 1011 + 0001011 + 000001011.$$

We begin in state q_0 with x and y each written on one of the input tapes (tapes 1 and 2) in reverse binary notation, and all three reader heads lined up. We first add a terminal 0 to the end of y, which is necessary to ensure that our computation will halt.

Suppose there are k initial 0's in y. For each such 0, we replace it with a #, write a 0 on tape 3 in the corresponding cell, move the heads above tapes 2 and 3 one cell to the right, and stay in state q_0.

When we encounter the first 1 of y, we replace this 1 with a # and transition to a state q_C in which we copy x to tape 3 (the output tape), beginning in the ith cell of this tape. As the contents of tape 1 are being copied onto tape 3, we also require that head above tape 2 moves in the same directions as the heads above tapes one and three until we encounter the first # on tape one. (One can verify that the heads above tapes 2 and 3 will always be lined up). We then transition to a state in which we reset the position of the tape 1 head to scanning the first bit of x, and then the tape 2 head is scanning the leftmost bit in y that has not been erased (i.e., replaced with a #). As above, we require that the tape 3 head moves in the same directions as the tape 2 head during this phase (which will ensure that we begin copying x in the correct place on the tape if the next bit of y happens to be a 1). We then transition back to state q_0 and continue as before.

Suppose that the next bit of y that is equal to 1 is the ith bit of y. Then we start the addition by adding the first bit of x to the ith bit of the output tape. We note that we may need to utilize a carry state during the addition. When we reach the # at the end of y, the computation is complete, and we transition to the halting state.

Like finite state machines, Turing machines can be represented by a state diagram.

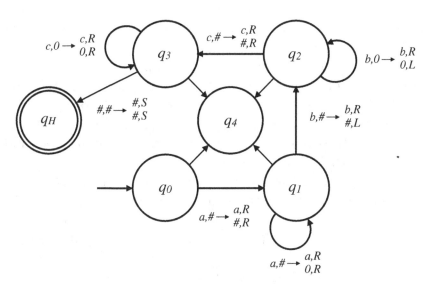

Example 6.2.4. There is a Turing machine that accepts the set $\{a^n b^n c^n : n > 0\}$ over the alphabet $\Sigma = \{a, b, c, 0, \#\}$.

Here we will use the "0" as a special marking symbol, although we can do without it (and use "a" instead). The state diagram above gives a Turing machine which accepts the above set.

Observe that q_4 functions as a reject state. Any input not listed on any of the edges will cause a transition to state q_4.

Example 6.2.5. Here is a sketch for a Turing machine M which accepts the set of propositional sentences over $\mathcal{L} = \{A, \neg\}$, that is,

$$L(M) = \{A, (\neg A), (\neg(\neg A)), \ldots\}.$$

M has the input tape and a scratch tape where a count is kept of the stack of $\neg$'s that have been seen.

The initial state q_0 transitions to the halting state q_H on input A and to the state q_1 on input the symbol (while moving right. State q_1 transitions to q_2 on input $\neg$, again moving right, and writes a 1 on the scratch tape. These states are not influenced by the contents of the scratch tape.

State q_2 transitions to state q_1 on input the symbol (and moves right on both tapes. State q_2 transitions to state q_4 on input A with 1 on the scratch tape, moves right on the input tape, and does nothing on the scratch tape. State q_3 transitions back to state q_2 on input $\neg$, moves right on input tape, and writes 1 on the scratch tape.

State q_4 remains in state q_4 when reading the symbol) on the input tape and 1 on the scratch tape, it moves right on the input tape, deletes 1 and moves left on the scratch tape. State q_4 transitions to the halting state when reading blanks on both tapes.

All other possibilities lead to rejection, such as input $\neg$ in state q_2.

Consider the action of M on input $(\neg(\neg A))$. M goes through states $q_0, q_1, q_2, q_1, q_2, q_4, q_4, q_4, q_H$.

Example 6.2.6. We define a Turing machine M that computes the function $f(x) = |x|$ (the length of a string x). There are two tapes, the input tape and the output tape. The input tape is read-only but we allow writing on the output tape. Let the input alphabet be $\Sigma = \{a\}$. Let α/β indicate that the machine is currently reading α on the input tape and β on the output tape. Similarly D_1/D_2 indicates that the head on the input tape moves in direction D_1 while the head on the output tape moves in direction D_2. The idea is to add one to the output tape after reading each symbol of the input tape. State q_1 arises when we need to add one to the output by carrying. State q_2 simply brings the output tape back to the first bit. Certain transitions are omitted from the table since they lead to a divergent computation (we will assume that incorrect inputs will immediately cause a transition to a reject state). For example, we only get to state q_1 when we have read a on the first tape and we continue to read that a, so that the input $\#/0$ is not a legal input when we have reached state q_1.

The following table describes the behavior of the three main states.

state	read	write	move	new state
q_0	$a/\#$	$a/1$	R/S	q_0
q_0	$a/0$	$a/1$	R/S	q_0
q_0	$a/1$	$a/0$	S/R	q_1
q_0	$\#/0$	$\#/0$	S/S	q_H
q_0	$\#/1$	$\#/1$	S/S	q_H
q_0	$\#/\#$	$\#/0$	S/S	q_H
q_1	$a/\#$	$a/1$	S/L	q_2
q_1	$a/0$	$a/1$	S/L	q_2
q_1	$a/1$	$a/0$	S/R	q_1
q_2	$a/\#$	$a/\#$	R/R	q_0
q_2	$a/0$	$a/0$	S/L	q_2
q_2	$a/1$	$a/1$	S/L	q_2

We now prove some general facts about certain kinds of languages that are central to the study of Turing machines.

Definition 6.2.7. Fix a finite alphabet Σ.

(a) A partial function $f : (\Sigma^*)^k \to \Sigma^*$ is *Turing computable* if $f = f_M$ for some Turing machine M.
(b) A language L is *Turing computable* if the characteristic function of L is Turing computable.
(c) A language L is said to be *Turing semicomputable* if there is a Turing machine M such that $L = \{w : M(w)\!\downarrow\}$.

Example 6.2.8. Here is a simple Turing machine M such that $M(w)\!\downarrow$ if and only if w contains a 0. In state q_0, M moves right and remains in state q_0 upon reading a 1 or a blank. M immediately halts upon reading a 0.

Proposition 6.2.9.

(a) *Every Turing computable language is also Turing semicomputable.*

(b) *A language L is Turing computable if and only if both L and its complement are Turing semicomputable.*

Proof.   (a) Let M be a Turing machine that computes the characteristic function of L. We modify M to define a machine M' as follows. First we introduce new states q_A and q_B. Replace any transition that goes to the halting state q_H with a transition that goes the the state q_A. For the q_A state, add two transitions. If the output tape reads 1, then transition to the halting state q_H. If the output tape reads 0, then move the output tape head one cell to the right and transition to state q_B. In state q_B, move the output tape head one cell to the left and return to state q_A. Then $M'(w)$ will halt if and only if $M(w) = 1$ and will endlessly alternate between states q_A and q_B if and only if $M(w) = 0$.

   (b) First observe that if $L \subseteq \Sigma^*$ is Turing computable, then $\Sigma^* \setminus L$ is Turing computable. Indeed, if M computes the characteristic function of L, then define M' to be the machine that behaves exactly like M except that for $i = 0, 1$ whenever M writes i on its output tape, M' writes $1 - i$ on its output tape. It follows from part (a) that if L is Turing computable, then both L and its complement are semicomputable.

   Now suppose that $L = \{w : M_0(w){\downarrow}\}$ and that $\Sigma^* \setminus L = \{w : M_1(w){\downarrow}\}$ for two Turing machines M_0 and M_1. We define a Turing machine M such that the function f_M computed by M is the characteristic function of L. Suppose for the sake of simplicity that M_0 and M_1 each have one input tape and have no output tape. Then M will have one input tape, two scratch tapes, and one output tape. The states of M will include pairs (q, q') where q is a state of M_0 and q' is a state of M_1. Given w on the input tape of M, M will begin by copying w onto each of the scratch tapes and transitioning to the pair (q_0, q'_0) of initial states of M_0 and M_1. On the first scratch tape, M will simulate M_0, while on the second scratch tape, M will simulate M_1. Eventually M will enter a state of one of the two forms: (q_H, r), where q_H is the halting state of M_0 and r is a non-halting state of M_1, or (q, q'_H), where q'_H is the halting state of M_1 and q is a non-halting state of M_0.

- If M enters a state of the form (q_H, r), this means that the machine M_0 has halted on w, and hence $w \in L$. M will thus write a 1 on its output tape and transition to its halting state.
- If M enters a state of the form (q, q'_H), this means that the machine M_1 has halted on w, and hence $w \notin L$. M will thus write a 0 on its output tape and transition to its halting state.

Note that M will never enter the state (q_H, q'_H), since M_0 and M_1 can never accept the same word. It thus follows that M computes the characteristic function of L. □

For more about automata and languages, see [20] or [29].

We conclude this section with a discussion of universal Turing machines. First we introduce some useful notation.

Definition 6.2.10. Given two Turing machines M_0 and M_1, we write $M_0(x) \simeq M_1(x)$ to mean that (i) $M_0(x)\!\downarrow$ if and only if $M_1(x)\!\downarrow$ and (ii) $M_0(x)\!\downarrow$ implies that $M_0(x) = M_1(x)$.

Theorem 6.2.11. *Fix a finite alphabet* $\Sigma = \{0, 1, q, \#, *, L, R, S, H, \downarrow\}$. *There is a* universal *Turing machine* $U : \Sigma^* \times \Sigma^* \to \Sigma^*$ *such that, for any Turing machine* $M : \{0,1\}^* \to \{0,1\}^*$, *there exists* $w_M \in \Sigma^*$ *such that, for all inputs* $x \in \{0,1\}^*$, $U(w_M, x) \simeq M(x)$.

We will only sketch the main ideas in the proof of Theorem 6.2.11. The extra symbols $\#$ and $*$ are used as markers to locate and separate different parts of the tapes. To simplify matters, we will use the following lemma.

Lemma 6.2.12. *Let M be a k-tape Turing machine for some $k > 1$. Then there is a single-tape Turing machine M' such that $M(x) \simeq M'(x)$ for all $x \in \Sigma^*$.*

Before beginning the proof, we want to consider further the notion of a Turing machine configuration. Suppose that M is a single-tape Turing machine working on the alphabet $\Sigma = \{0, 1\}$ with states $q_0, q_1, \ldots, q_k$. The computation of $M(w)$ on an input w is accomplished in a series of steps which may be given by the sequence of configurations it goes through. The configuration $ua * q_i \downarrow bv$ indicates that M is in state q_i, that the word $uabv$ is written on the tape and that the head is located at the symbol b; here u and v are words in the language $\{0, 1\}^*$. The *next* configuration of M in the computation may be determined by looking at the transition $\delta(q_i, b)$.

Proof of Theorem 6.2.11. (Sketch) We define a universal Turing machine with two input tapes, a scratch tape, and an output tape. For each machine M, we would like to define a word $w_M \in \Sigma^*$ that encodes all of the information of M. We will let w_M be the entire transition table of M written as one long string in the alphabet Σ^*, with a $*$ separating each entry on a given row of a table and $**$ separating each row of the table. The string $q0$ will stand for the initial state, qH will stand for the halting state, and all other states will be coded by a q followed by a finite string of 1s ($q1$, $q11$, $q111$, etc).

Now, given the code w_M for a machine M, to compute $U(w_M, x)$ (i.e., $M(x)$), U proceeds by writing the initial configuration of M with input x on its scratch tape. Suppose, for example that $x = 000$. Then U will write

$$q0 \downarrow 000,$$

where the $\downarrow$ specifies that the reader head is above the first 0 on the input tape of M. To proceed, U simply consults the transition table w_M written on its first input tape and writes the resulting configuration of M after one move on its scratch tape. Continuing the example from above, if the first move of M is to change to state q_1, replace the first 0 of x with a 1 and move the head one cell to the right, the resulting configuration

will be
$$1 * q1 \downarrow 00.$$

Continuing in this way, if $M(x)\downarrow$, then U will eventually come to a stage in which a halting configuration is written on its scratch tape. In this halting configuration, the value $y = M(x)$ will be written, and so U can simply copy this value y to its output tape and transition to its own halting state. Lastly, if $M(x)\uparrow$, then $U(w_M, x)\uparrow$. □

It is important to note that our use of the alphabet Σ in the definition of a universal Turing machine is not strictly necessary. For instance, we can also define a universal Turing machine $U : \{0,1\}^* \times \{0,1\}^* \to \{0,1\}^*$ by representing each of the symbols in Σ as a unique binary string. Some caution is necessary to make sure the coding is unambiguous. We will not discuss the details of such a coding here, but we will assume that for each Turing machine M, there is some unique $w_M \in \{0,1\}^*$ that codes M. Moreover, we will assume that every $x \in \{0,1\}^*$ codes some Turing machine, which we will write as M_x. We will justify these assertions in the next section when we discuss Gödel numbering of Turing machines.

Exercises for Section 6.2

Exercise 6.2.1. Describe a Turing machine M such that $M(w) = ww$ for any word $w \in \{a, b\}^*$. (For example, $M(aab) = aabaab$).

Exercise 6.2.2. Describe a Turing machine M such that $M(u, v)$ halts exactly when $u = v$ for $u, v \in \{a, b\}^*$. (M begins with u on one input tape and with v on a second input tape. There might be a scratchwork tape).

Exercise 6.2.3. Describe a Turing machine M which halts on input (u, v) if and only if $u \leq v$ in the standard dictionary ordering on words from the alphabet $\{a, b, c\}$.

Exercise 6.2.4. Describe two Turing machines M and N such that for any natural number $n = \Sigma_{i \leq k} a_i 2^{-i-1}$,

- $f_M(1^n) = a_0 a_1 \cdots a_k$ and
- $f_N(a_0 a_1 \cdots a_k) = 1^n$.

6.3.   Recursive Functions

In this section, we define the primitive recursive and the (partial) recursive functions and show that they are all Turing computable. Each function f maps from $\mathbb{N}^k$ to $\mathbb{N}$ for some fixed k (the *arity* of f). We note that the term "computable" has replaced the word "recursive" in a large part of the modern literature.

We first define the notions of composition and primitive recursion.

Definition 6.3.1.

(1) The operation of *composition* of functions $f : \mathbb{N}^k \to \mathbb{N}$ and $g_i : \mathbb{N}^j \to \mathbb{N}$ for $i = 1, \ldots, k$ (where $j, k > 0$) that are in $\mathcal{F}$, produces the function $h : \mathbb{N}^j \to \mathbb{N}$, defined by

$$h(x_1, \ldots, x_j) = f(g_1(x_1, \ldots, x_j), \ldots, g_k(x_1, \ldots, x_j)).$$

(2) The operation of *primitive recursion* on two functions $g : \mathbb{N}^k \to \mathbb{N}$ and $h : \mathbb{N}^{k+2} \to \mathbb{N}$ produces the function $f : \mathbb{N}^{k+1} \to \mathbb{N}$, defined by

$$f(0, x_1, \ldots, x_k) = g(x_1, \ldots, x_k),$$
$$f(n + 1, x_1, \ldots, x_k) = h(n, x_1, \ldots, x_k, f(n, x_1, \ldots, x_k)).$$

Example 6.3.2. The addition function $f(x, y) = x + y$ may be defined by primitive recursion as follows:

$$f(0, y) = g(y) = y,$$
$$f(n + 1, y) = h(n, y, f(n, y)) = f(n, y) + 1.$$

Definition 6.3.3. The collection of *primitive recursive functions* is the smallest collection of functions from $\mathbb{N}^k$ to $\mathbb{N}$ for each $k > 0$ that includes the following *initial functions*:

(1) the constant function $c(x) = 0$,
(2) the successor function $s(x) = x + 1$,
(3) the projection functions $p_i^k(x_1, \ldots, x_k) = x_i$ for each $k \in \mathbb{N}$ and $i = 1, \ldots, k$,

and is closed under

(4) composition and
(5) primitive recursion.

Example 6.3.4.

(1) For any constant $c \in \mathbb{N}$, the function $h(x) = c$ is primitive recursive. The proof is by induction on c. For $c = 0$, this follows from the fact that the initial function $c(x) = 0$ is primitive recursive. Supposing that $g(x) = c$ is primitive recursive and using the fact that $s(x) = x + 1$ is primitive recursive, we can use composition to conclude that $h(x) = s(g(x)) = g(x) + 1 = c + 1$ is primitive recursive.
(2) For any k and any c, the constant function $f : \mathbb{N}^k \to \mathbb{N}$ with

$$f(x_1, \ldots, x_k) = c$$

is primitive recursive. We have $h(x) = c$ by (1) and we have

$$p_1^k(x_1, \ldots, x_k) = x_1$$

as a basic function, so that we have the primitive recursive definition $f(x_1, \ldots, x_k) = h(p_1^k(x_1, \ldots, x_k)) = h(x_1) = c$.
(3) The addition function $f(x, y) = x + y$ is primitive recursive. Let $g(y) = p_1^1(y)$ and $h(x, y, z) = c(p_3^3(x, y, z))$. Then f is given by the primitive recursive scheme

$$f(0, y) = g(y) = y,$$
$$f(n + 1, y) = h(n, y, f(n, y)) = f(n, y) + 1.$$

(4) The *predecessor* function $f(x) = x \dotdiv 1$, defined by

$$f(x) = \begin{cases} x - 1 & \text{if } x > 0, \\ 0 & \text{otherwise,} \end{cases}$$

is primitive recursive. Let $g(x) = c(x)$ and $h(x, y) = p_1^2(x, y)$. Then f is given by the primitive recursive scheme

$$f(0) = g(y) = 0,$$
$$f(n + 1) = h(n, f(n)) = n.$$

(5) The *truncated subtraction* function

$$f(x, y) = \begin{cases} y \dotdiv x & \text{if } x \le y, \\ 0 & \text{otherwise,} \end{cases}$$

is primitive recursive. Let $g(y) = p_1^1(y)$ and $h(x, y, z) = z \dotdiv 1$. Then f is given by the primitive recursive scheme

$$f(0, y) = g(y) = y,$$
$$f(n + 1, y) = h(n, y, f(n, y)) = f(n, y) \dotdiv 1.$$

(6) The multiplication function $f(x, y) = x \cdot y$ is primitive recursive. Let $g(x) = c(x) = 0$ and let $h(x, y, z) = p_3^3(x, y, z) + p_2^3(x, y, z)$. Then f is given by the primitive recursive scheme

$$f(0, y) = g(y) = 0,$$
$$f(n + 1, y) = h(n, y, f(n, y)) = f(n, y) + y.$$

Definition 6.3.5. A relation $R \subseteq \mathbb{N}^k$ is primitive recursive if the characteristic function χ_R is primitive recursive.

Example 6.3.6. The set of even numbers is primitive recursive since its characteristic function may be defined by the scheme $f(0) = 1$ and $f(n + 1) = 1 - f(n)$.

Example 6.3.7. Observe that for any natural numbers x and y, $x \leq y$ if and only if $x \dot{-} y = 0$, so that the order relation is primitive recursive. It is easy to see that the strict order is also primitive recursive and that equality is primitive recursive.

Definition 6.3.8. The function $h : \mathbb{N}^k \to \mathbb{N}$ is *defined by cases* from two functions $f : \mathbb{N}^k \to \mathbb{N}$ and $g : \mathbb{N}^k \to \mathbb{N}$ and a relation $R \subseteq \mathbb{N}^k$ if, for all $x_1, \ldots, x_k$,

$$h(x_1, \ldots, x_k) = \begin{cases} f(x_1, \ldots, x_k) & \text{if } R(x_1, \ldots, x_k), \\ g(x_1, \ldots, x_k) & \text{otherwise.} \end{cases}$$

Example 6.3.9. A *piecewise* linear function can be defined on $\mathbb{N}$ by

$$h(x) = \begin{cases} 3x & \text{if } x < 5, \\ 2x + 5 & \text{if } x \geq 5. \end{cases}$$

Using the examples above, we can now show the following proposition.

Proposition 6.3.10. *If the function h is defined by cases from two primitive recursive functions f and g and a primitive recursive relation R, then h is also primitive recursive.*

Proof. Given the definition of h as above from f, g and R, we have, for all $\vec{x} = (x_1, \ldots, x_k)$,

$$h(\vec{x}) = \chi_R(\vec{x}) \cdot f(\vec{x}) + (1 - \chi_R(\vec{x})) \cdot g(\vec{x}).$$

Example 6.3.11. The *sign* function sg, defined by

$$sg(x) = \begin{cases} 0 & \text{if } x = 0, \\ 1 & \text{if } x > 0, \end{cases}$$

is primitive recursive by Proposition 6.3.10.

Note that every primitive recursive function $f : \mathbb{N}^k \to \mathbb{N}$ is total, i.e., defined for very input in $\mathbb{N}^k$. This clear from the fact that composition and primitive recursion operations applied to total functions return a total function. We can extend the collection of primitive recursive functions to the collection of partial recursive functions by adding one additional scheme for defining new functions from previously defined ones. This scheme allows for the possibility that a function be undefined on a given input. Given a function $f : \mathbb{N}^k \to \mathbb{N}$, if f is defined on input $(x_1, \ldots, x_k)$, we write $f(x_1, \ldots, x_k)\downarrow$; if f is undefined on $(x_1, \ldots, x_k)$, we write $f(x_1, \ldots, x_k)\uparrow$.

Definition 6.3.12. The operation of *unbounded search* on a (possibly partial) function $g : \mathbb{N}^{k+1} \to \mathbb{N}$ produces the (possibly partial) function $f : \mathbb{N}^k \to \mathbb{N}$ defined to be the least n such that $g(n, x_1, \ldots, x_k) = 0$ and $g(i, x_1, \ldots, x_k)\downarrow$ for $i \le n$, if such n exists, and $f(x_1, \ldots, x_k)\uparrow$, otherwise. The operation of *bounded search* on a function g produces a function f such that $f(n, x)$ equals the least $i < n$ such that $g(i, x) = 0$, if any such n exists, and otherwise $f(n, x) = n$.

Example 6.3.13. The function $f(x) = \lfloor x/2 \rfloor$ (the integer part of $x/2$) may be defined as

$$f(x) = (\text{the least } n) \; x \dot- (n + n + 1) = 0.$$

The partial function $f(x) = x/2$, defined only on even numbers, has a similar definition.

Bounded notions of search and also quantification are important. We note that if we add a scheme of *bounded search* to the collection of primitive recursive functions, we do not add any new functions.

Proposition 6.3.14. *If g is primitive recursive and f is defined from g by bounded search, then f is primitive recursive.*

Proof. We have $f(0, x) = 0$ and for each n,

$$f(n + 1, x) = \begin{cases} f(n, x) & \text{if } g(f(n, x), x) = 0, \\ n + 1 & \text{otherwise.} \end{cases}$$ □

Proposition 6.3.15. *If R and S are primitive recursive relations, then $\neg R$, $R \vee S$, and $R \wedge S$ are all primitive recursive relations.*

The proof is left as an exercise.

Definition 6.3.16. The relation R is defined from S by bounded existential quantification if for all $x_1, \ldots, x_n$:

$$R(y, x_1, \ldots, x_n) \iff (\exists z < y) S(z, y, x_1, \ldots, x_n).$$

Bounded universal quantification is similarly defined.

Proposition 6.3.17. *If S is primitive recursive and R is defined from S by bounded quantification, then R is primitive recursive.*

Proof. Let $P(t, y, \vec{x}) \iff (\exists z < t) S(z, y, \vec{x})$. Then $R(y, \vec{x}) \iff P(y, y, \vec{x})$ and P may be defined by primitive recursion since $P(0, y, \vec{x})$ is always false and $P(t+1, y, \vec{x})$ if and only if $P(t, y, \vec{x})$ or $S(t, y, \vec{x})$.

For universal quantification, we have

$$(\forall z < y) S(z, y, x_1, \ldots, x_n) \iff \neg(\exists z < y) \neg S(z, y, x_1, \ldots, x_n).$$ □

We can now use bounded search and quantification to obtain some primitive recursive sets and functions.

Example 6.3.18. Define the functions Q and R as follows. For $b > 0$, let $Q(a, b)$ be the quotient when a is divided by b and let $R(a, b)$ be the remainder, so that $a = Q(a, b) \cdot b + R(a, b)$

(to ensure that Q and R are total, we can set $Q(a,0) = 0$ and $R(a,0) = a$ for every a). That is, for $b > 0$, $Q(a,b)$ is the least $i \leq a$ such that $(i+1) \cdot b > a$ and $R(a,b) = a - Q(a,b) \cdot b$. Then both Q and R are primitive recursive.

Example 6.3.19. The relation $x \mid y$ (x divides y) is primitive recursive, since $x \mid y$ if and only if $(\exists q < y + 1)\, x \cdot q = y$.

Example 6.3.20. The set of prime numbers is primitive recursive and the function P which enumerates the prime numbers in increasing order is also primitive recursive. To see this, note that $p > 1$ is prime if and only if $(\forall x < p)[x \mid p \rightarrow x = 1]$. Now we know that for any prime p, there is another prime $q > p$ with $q < p! + 1$. By Exercise 6.3.4, the factorial function is primitive recursive. Then we can recursively define P by $P(0) = 2$ and, for all i,

$$P(i+1) = (\text{the least } x < P(i)! + 1) \ x \text{ is prime}.$$

Definition 6.3.21. The family of Δ_0 formulas in the language $\{+, \cdot, S, 0, <\}$ of arithmetic is the smallest collection of formulas including all atomic formulas and closed under the propositional connectives and bounded quantification.

Proposition 6.3.22. *For every Δ_0 formula $\phi(x_1, \ldots, x_k)$, the set $\{\vec{x} : \mathbb{N} \models \phi(\vec{x})\}$ is a primitive recursive relation.*

Proof. The proof is a straightforward induction on rank over the set of Δ_0 formulas. We have seen in the examples above that this holds for the atomic formulas. Proposition 6.3.15 provides the closure under propositional connectives and Proposition 6.3.17 gives the closure under bounded quantification. □

Definition 6.3.23. The collection of *partial recursive* (*or partial computable*) *functions* is the smallest collection $\mathcal{F}$ of functions from $\mathbb{N}^k$ to $\mathbb{N}$ for each $k > 0$ that includes the primitive recursive functions and is closed under composition, primitive recursion, and

(6) unbounded search.

We will refer to each total recursive function simply as a *recursive function*.

We now show that the collection of partial recursive functions is equivalent to the collection of Turing computable functions, in the sense that every partial recursive function can be computed by a Turing machine and every Turing computable function is partial recursive.

Theorem 6.3.24. *Every partial recursive function can be computed by a Turing machine.*

Proof. The proof is by induction on the family of partial recursive functions. For the base case, it is clear that the initial functions are Turing computable. We now verify that the schemes of composition, primitive recursion, and unbounded search yield Turing computable functions when applied to Turing computable functions.

Composition. For simplicity let $f(x) = h(g(x))$ where h is computable by machine M_h and g is computable by machine M_g. Assume without loss of generality that each machine uses one input tape and one output tape and that the sets of states of the two machines are disjoint. We define a machine M_f that computes f with four tapes:

(1) the input tape;
(2) a scratch tape to serve as the output tape of g;
(3) a scratch tape to serve as the input tape for h; and
(4) an output tape.

M will have the states of g and h together plus a few new states to handle the transfer from M_g to M_h. The transition function for M_g will be changed so that instead of halting when the output $g(x)$ is ready, M will go into a subroutine which copies from the g-output tape to the h-input tape and then hands over the controls to M_h. The halting states of M will be the halting states of M_h.

Primitive Recursion. For simplicity let $f(0, x) = g(x)$ and $f(n + 1, x) = h(n, x, f(n, x))$. Let M_g compute g and let M_h compute h. Assume without loss of generality that M_g has one read-only input tape and one output tape and that M_h has three read-only input tapes and one output tape, and that the sets of states of the two machines are disjoint. We define a machine M that computes $f(n, x)$ with seven tapes:

(1) the input tape for n;
(2) the input tape for x;
(3) a tape to keep track of the ongoing value of $m < n$;
(4) a tape to keep track of the ongoing value of $f(m, x)$;
(5) a scratch tape to serve as the input tape for h;
(6) a scratch tape to serve as the output tape of $h(m, x, f(m, x))$; and
(7) an output tape.

M will have the states of g and h together plus a few new states to handle the transfer from M_g to M_h and the ongoing recursion. First, M will check if the input written on tape (1) is a 0; if so, it will simulate M_g on the input on tape (2) and provide this as output on tape (7) and halt. Otherwise, M will use tape (4) for the output of M_g and the transition function for M_g will be changed so that instead of halting when the output $g(x)$ is ready, M write $m = 0$ onto tape (3), and then hand over control to M_h. The inputs for M_h are found on tapes (2), (3) and (4). M_h uses these to compute $h(m, x, f(m, x)) = f(m+1, x)$. When M_h is ready to halt and give its output, M does the following:

(i) M compares m from tape (5) with n from tape (1); if $n = m+1$, then the value on tape (6) equals the desired $f(n, x)$, so M copies this to tape (7) and halts.

(ii) Otherwise, M erases tape (4) and then copies the value from tape (6) onto tape (4).

(iii) Then M adds one to the value of m on tape (3), erases tape (6) and hands control back to M_h again.

Unbounded Search. For simplicity let $f(x) =$ the least n such that $g(n, x)\downarrow = 0$ and for all $i \leq n$ $g(i, x)\downarrow$. Let M_g compute g using one input tape and one output tape.

We define a machine M that computes $f(x)$ with five tapes:

(1) the input tape for x;
(2) a tape for the ongoing value of n;
(3) a scratch tape to serve as the input tape for g;
(4) a tape to keep track of the ongoing value of $g(n, x)$; and
(5) an output tape.

M will have the states of g plus a few new states to handle the the ongoing computations of $g(n, x)$. M begins by writing $n = 0$ on tape (2) and handing control to M_g. The transition function for M_g will be changed so that instead of halting when the output $g(n, x)$ is ready, M will do the following:

(i) Compare the value $g(n, x)$ from tape (4) with 0. If $g(n, x) = 0$, then the value n on tape (2) equals the desired $f(x)$, so M copies this to tape (5) and halts.

(ii) Otherwise, M increments the value of n on tape (2), erases tapes (3) and (4) and hands control back to M_g again. $\square$

We now turn to the other direction, namely, that every Turing computable function is partial recursive. We first consider the problem of coding words on a finite alphabet $\Sigma = \{a_1, \ldots, a_k\}$ into words over $\{0, 1\}^*$ and then into natural numbers. For the first part, we can code the word $w = a_{i_1} a_{i_2} \cdots a_{i_n}$ into the string $0^{i_1} 1 0^{i_2} 1 \cdots 0^{i_n} 1$. For the second part, we can code the word $v = i_0 \cdots i_n \in \{0, 1\}^*$ by the reverse binary natural number $i_0 \cdots i_n 1$. Note that this will work for an arbitrary finite alphabet and even for the potentially infinite alphabet $\mathbb{N}$. Let $\langle w \rangle$ be the natural number code for the word w.

It is clear that we can use a Turing machine to compute the code $\langle w \rangle$ for a string $w \in \{a_1, a_2, \ldots, a_k\}^*$: when reading each symbol a_i, the machine moves through a sequence of states that results in it writing i 0's followed by a 1 and returning to its initial state.

For the other direction, given a number n written in reverse binary notation as $0^{r_0} 1 0^{r_1} 1 \cdots 0^{r_k} 1$, we can compute a sequence $n_0, r_0, n_1, r_1, \ldots, n_k, r_k$ with $n = n_0$ as follows. Recall the primitive recursive functions $Q(a, b)$, the quotient when a is divided by b and the corresponding remainder $R(a, b)$, and note that b divides a if and only if the remainder is 0. Now let r_0 be the least r such that $R(n_0, 2^{r+1}) \neq 0$ (i.e., 2^{r+1} does not divide n_0) and then let n_1 satisfy $2 \cdot n_1 = Q(n_0, 2^{r_0}) - 1$. After this, let r_{i+1} be the least r such that $R(n_i, 2^{r+1}) \neq 0$ and let n_{i+1} satisfy $2 \cdot n_{i+1} = Q(n_i, 2^{r_i}) - 1$. Eventually this process will terminate with $n_k = 0$ for some k. Thus the function which computes r_i from n and i is computable.

For example, if $n = 0^3 1 0^5 1 0 1$, then $r_0 = 3, n_1 = 0^5 1 0 1$, $r_1 = 5, n_2 = 01, r_2 = 1, n_3 = 0$.

Since we have algorithms to compute the natural number code $\langle w \rangle$ of a string w and also algorithms to compute the string w from the natural number $k = \langle w \rangle$, we can say that a function $F : \Sigma^* \to \Sigma^*$ is computable if and only if the function $f : \mathbb{N} \to \mathbb{N}$ is computable, where

$$f(\langle w \rangle) = \langle F(w) \rangle,$$

and similarly for functions with several variables.

Then we may use the universal Turing machine U to provide an enumeration $\{M_e : e \in \omega\}$ of all Turing computable functions by letting M_e be the machine whose program may be coded by the natural number e. Then the proof of Theorem 6.2.11 also proves the following.

Proposition 6.3.25. *For natural numbers s, e, w, let $F(s, e, w)$ be the natural number which codes the sth configuration in the computation of M_e on input word coded by w. Let $M_e^s(w)$ be the output given by M_e if the computation halts by stage s. Then*

(1) *the function F is primitive recursive;*
(2) *the set $\{(e, s, w) : M_e^s(w){\downarrow}\}$ is primitive recursive;*
(3) *the set $\{(e, s, w, v) : M_e^s(w) = v\}$ is primitive recursive.*

Theorem 6.3.26. *Every Turing computable function is a partial recursive function.*

Proof. Given any Turing computable function M_e, using Proposition 6.3.25, $M_e(w)$ may be computed by searching for the least s such that $M_e^s(w){\downarrow}$ and then outputting the unique $v \le s$ such that $M_e^s(w) = v$.

Since we can code finite sequences as strings and hence as natural numbers, the equivalence of Turing computable functions from $\mathbb{N}^k \to \mathbb{N}$ easily follows. This completes the proof that any Turing computable function is partial recursive. □

Exercises for Section 6.3

Exercise 6.3.1. Show that if R and S are recursive relations, then $\neg R$, $R \vee S$, and $R \wedge S$ are all recursive relations, and similarly for primitive recursive relations.

Exercise 6.3.2. Show that function $F(a, b) = \mathrm{LCM}(a, b)$ is primitive recursive.

Exercise 6.3.3. Show that the general exponentiation function $f(x, y) = y^x$ is primitive recursive.

Exercise 6.3.4. Suppose that f is a (total) recursive (or primitive recursive) function and let $g(x, y) = \Pi_{i < y} f(x, i)$. Show carefully that g is also recursive (primitive recursive).

Use this to show that the function $f(n) = n!$ is primitive recursive.

Exercise 6.3.5. Suppose that $f : \mathbb{N} \to \mathbb{N}$ is a (total) recursive one-to-one function. Show that f^{-1} is also recursive.

6.4.   The Halting Problem

In this section, we describe a set of natural numbers known as the Halting Problem and show that it is *unsolvable*, that is, not computable. The notion of a computably enumerable set will be needed.

Definition 6.4.1. A subset A of Σ^* is said to be *computably enumerable* if there is some computable function $f : \mathbb{N} \to \Sigma^*$ such that $A = \text{ran}(f)$. That is, A can be enumerated as $f(0)$, $f(1), f(2), \ldots$.

Proposition 6.4.2. *Let $B \subseteq \Sigma^*$ and let $A = \{\langle w \rangle : w \in B\}$. Then B is computable if and only if A is computable and similarly for semicomputable and computably enumerable.*

Proof.   We saw above that the coding and decoding functions mapping the string w to the natural number $\langle w \rangle$ and mapping $\langle w \rangle$ back to w are both primitive recursive. Thus for any function $F : \Sigma^* \to \Sigma^*$, F is computable if and only if the function $f : \mathbb{N} \to \mathbb{N}$, defined by $f(\langle w \rangle) = \langle F(w) \rangle$ is computable.

Now suppose that $B \subseteq \Sigma^*$ is semicomputable, let $B = \text{dom}(F)$, and define f so that $f(\langle w \rangle) = \langle F(w) \rangle$. Then $A = \text{dom}(f)$ and is semicomputable.

Conversely, suppose that $A = \{\langle w \rangle : w \in B\} = \text{dom}(f)$ is semicomputable and let $F(w) = f(\langle w \rangle)$. Then $B = \text{dom}(F)$.

Next suppose that B is computably enumerable, let $B = \text{ran}(F)$, where $F : \mathbb{N} \to \Sigma^*$, and let $f(n) = \langle F(n) \rangle$. Then clearly $A = \text{ran}(f)$ and is computably enumerable.

Conversely, suppose that $A = \{\langle w \rangle : w \in B\} = \text{ran}(f)$ is computably enumerable and let $F(n)$ be the string coded by $f(n)$. Then $B = \text{ran}(F)$ and is computably enumerable. □

Theorem 6.4.3. *A is computably enumerable if and only if A is semicomputable.*

Proof. By Proposition 6.4.2, it suffices to show this for $A \subseteq \mathbb{N}$. Suppose first that A is computably enumerable. If $A = \emptyset$, then certainly A is semicomputable, so we may assume that $A = \text{rng}(f)$ for some computable function f. Now define the partial computable function ϕ by $\phi(x) = (\text{least } n)f(n) = x$ for $x \in \mathbb{N}$. Then $A = \text{dom}(\phi)$, so that A is semicomputable.

Next suppose that $A \subseteq \mathbb{N}$ is semicomputable and let ϕ be a partial computable function so that $A = \text{dom}(\phi)$. If A is empty, then it is computably enumerable. If not, select $a \in A$ and define the computable function f by

$$f(2^s \cdot (2m+1)) = \begin{cases} m & \text{if } \phi(m)\!\downarrow \text{ in } < s \text{ steps,} \\ a & \text{otherwise.} \end{cases}$$

Then $A = \text{rng}(f)$, so that A is computably enumerable. □

Proposition 6.4.4. *For any partial function $\varphi : (\Sigma^*)^n \to \Sigma^*$, φ is partial computable if and only if*

$$\text{graph}(\varphi) = \{(x_1, \ldots, x_n, y) : \varphi(x_1, \ldots, x_n) = y\}$$

is a computably enumerable set.

Proof. Suppose first that φ is partial computable and define $\psi(x, y) = \varphi(x)$ if $\varphi(x) = y$ and $\psi(x, y) \uparrow$ otherwise. Then $\text{dom}(\psi) = \text{graph}(\varphi)$. For the other direction, suppose that $G = \text{graph}(\varphi) = \text{rng}(\psi) = \{(x_0, y_0), (x_1, y_1), \ldots\}$ is computably enumerable. Given input x, let $h(x) = (\text{least } n)x = p_1^2(\psi(n))$ and then $\phi(x) = p_2^2(h(n))$. □

Programs nearly always have bugs, so they may not do what we want them to do. The problem of determining whether a given Turing machine M halts on input string w is the *Halting Problem*. Alternatively, let us define the halting set to be $H = \{e : M_e(e)\downarrow\}$ (here we will take e to be the code of the relevant input string). This form of the Halting Problem was formulated by Kleene in 1952.

Observe that H is semicomputable, as it is the domain of the function $f(e) = M_e(e)$. Therefore by Theorem 6.4.3, H is also computably enumerable. By contrast, we have the following.

Theorem 6.4.5. *The Halting Problem is not computable.*

Proof. We will show that the complement of H is not semicomputable, so that by Proposition 6.2.9, H is not computable. Suppose by way of contradiction that there is a Turing machine N such that, for all $x \in \mathbb{N}$,

$$N(x)\downarrow \iff M_x(x)\uparrow.$$

This Turing machine N must have some code e. So for all x,

$$M_e(x)\downarrow \iff M_x(x)\uparrow.$$

The desired contradiction arises when we set $x = e$. $\qquad\square$

Thus there exists a set which is semicomputable but not computable. This result is essentially due to Turing, and was later formalized by Kleene. The basic papers may be found in [8], edited by Martin Davis.

Exercise for Section 6.4

Exercise 6.4.1. $S \subseteq \mathbb{N}^k$ is said to be Σ_1^0 if there is a computable relation R such that $S(x_1, \ldots, x_k) \iff (\exists y)R(y, x_1, \ldots, x_k)$. Prove that S is Σ_1^0 if and only if S is semicomputable.

Chapter 7

Decidable and Undecidable Theories

In this chapter, we bring together the two major strands that we have considered thus far: logical systems (and, in particular, their syntax and semantics) and computability theory. We will briefly discuss decidable and undecidable logical systems, but our primary focus will be decidable and undecidable first-order theories. The notion of decidability was one of the key ideas of 20th century mathematics, and was given impetus by the famous *Hilbert's Problems* [15] presented at the 1900 International Conference of Mathematics in Paris. The second problem was to prove that the axioms of arithmetic are consistent. This was shown to be unattainable by Gödel's Incompleteness Theorem, presented in Section 7.3 below. A related question posed by Hilbert and Ackermann in [17], known as the *Entscheidungsproblem*, asked for an algorithm for deciding the truth or falsity of any statement of predicate logic. This was answered in the negative by Church's Undecidability Theorem (Theorem 7.3.12), which is also proved in this chapter. These results as presented here make use of a number of results from computability theory as outlined in Chapter 5.

7.1.   Decidable vs. Undecidable Logical Systems

Let us consider some examples of decidable and undecidable logical systems.

Definition 7.1.1. A formal proof system is *decidable* if there is an effective procedure that, given a sentence φ, outputs 1 if φ is provable and outputs 0 if φ is not provable.

For a given logical system such as propositional logic or predicate logic, we say that the *decision problem* for this system is the problem of determining whether a given formula is logically valid. Moreover, for a decidable logical system, we say that its decision problem is *solvable*; similarly, for an undecidable logical system, we say that its decision problem is *unsolvable*.

Here are two examples of decidable systems. More will be considered in Section 7.2.

Example 7.1.2. The decision problem for propositional logic is solvable. The method of truth tables provides an algorithm for determining whether a given propositional formula φ is provable.

Example 7.1.3. Monadic predicate logic is first-order logic with only 1-place predicates such as $R(x), B(y)$, etc. The decision problem for monadic predicate logic is solvable, using the following key result:

If φ is a sentence of monadic predicate logic consisting of k distinct monadic predicates and r distinct variables, then if φ is satisfiable, it is satisfiable in a model of size at most $2^k \cdot r$.

Thus if φ is not provable, then there is a model of size at most $2^k \cdot r$ in which $\neg\varphi$ is satisfied. Hence to determine if a sentence of monadic predicate logic is provable, we just check to see whether φ is true in all models of cardinality less than some finite bound, which can be done mechanically. This result was first shown by Löwenheim [24].

We will see below in Section 7.3 that predicate logic is undecidable.

Example 7.1.4. Even the limited system of *dyadic first-order logic*, with only 2-place predicates such as $R(x, y), B(y, z)$, etc., may be shown to be undecidable. One can in fact show that if our language includes just *one* 2-place predicate, this is sufficient to create a collection of sentences for which the decision problem is unsolvable.

7.2. Decidable Theories

A theory T is said to be *decidable* if there is an effective procedure which, given a sentence φ in the language of T, decides whether $T \vdash \varphi$ or not. A structure $\mathcal{A}$ is said to be *decidable* if its theory is decidable in the language $\mathcal{L}(\mathcal{A})$ with names for elements of $\mathcal{A}$.

In this section we identify a number of sufficient conditions for a theory T to have an algorithm that enables us to determine the consequences of T. Then we will provide some examples of decidable theories, including the theory of infinity and the theory of dense linear orders without endpoints.

Let Γ be a finite set of $\mathcal{L}$-formulas in a first-order language $\mathcal{L}$, and φ an $\mathcal{L}$-formula. Recall that a formal proof $\Gamma \vdash \varphi$ is a finite sequence of $\mathcal{L}$-formulas $\psi_1, \ldots, \psi_n$, where φ is the formula ψ_n and each $\psi_i \in \Gamma$ or follows from some subcollection of $\{\psi_1, \ldots, \psi_{i-1}\}$ by one of the rules of inference. We will make use of the following, which can be proved by showing that each use of the rules of inference in the predicate calculus can effectively verified to hold.

Proposition 7.2.1. *Let $\mathcal{L}$ be a finite set of symbols for predicate logic. The set of formal proofs in $\mathcal{L}$ is computable and, furthermore, the set of triples (π, φ, θ) such that π is a proof of θ with a single premise φ, is also computable.*

Definition 7.2.2. A theory Δ is *effectively axiomatizable* if there is a computably enumerable set Γ such that the deductive closures of Δ and Γ are the same.

The next result shows that any the set of consequences of an effectively axiomatizable theory is computably enumerable. That is to say the set of consequences of a c.e. theory is c.e.

Proposition 7.2.3. *Let $\mathcal{L}$ be a computable language for predicate logic, and let Γ be a semicomputable set of formulas. Then $\{\theta : \Gamma \vdash \theta\}$ is semicomputable.*

Proof. Let $\Gamma = \{\gamma_0, \gamma_1, \dots\}$ where the function taking n to γ_n is computable and let φ_n be the conjunction $\gamma_0 \wedge \cdots \wedge \gamma_n$. Now let $F(\theta)$ search for the least code of a proof of θ with unique premise of the form γ_n, for some n. Then $\theta : \Gamma \vdash \theta\} = \mathrm{dom}(F)$. $\square$

Theorem 7.2.4. *If Γ is an effectively axiomatizable, complete $\mathcal{L}$-theory, then Γ is decidable.*

Proof. By Proposition 7.2.3, $\mathrm{Th}(\Gamma)$ is computably enumerable. Say $\mathrm{Th}(\Gamma)$ has computable enumeration $\{f(0), f(1), \dots\}$. Since Γ is complete, it follows that for any sentence φ, either the sentence or its negation is in $\mathrm{Th}(\Gamma)$. Now let $h(\varphi) = (\text{least } n)(f(n) = \varphi \vee f(n) = \neg\varphi)$. Then $\varphi \in \mathrm{Th}(\Gamma) \iff f(h(\varphi)) = \varphi$. $\square$

Recall from Section 3.5 that an effective procedure for quantifier elimination may be used to show that a relational theory Γ is decidable.

Proposition 7.2.5. *If a theory Γ in a finite relational language has an effective quantifier elimination procedure, then $\mathrm{Th}(\Gamma)$ is decidable.*

Another important example of an effectively axiomatizable, complete theory, is the theory *DLOWE* of dense linear orders without endpoints. Recall from Chapter 3 that any nonempty model of *DLOWE* is infinite and that any two countably infinite models of *DLOWE* are isomorphic. It follows from Corollary 3.5.7 and Theorem 7.2.4 that *DLOWE* is decidable.

Other examples of decidable theories are as follows:

(1) the theory of successor;
(2) Presburger arithmetic (arithmetic without multiplication);
(3) the theory of arithmetic with only multiplication and no addition;
(4) the theory of real closed fields (fields F in which every polynomial of odd degree has a root in F);
(5) the theory of algebraically closed fields of a fixed characteristic; and
(6) the theory of Euclidean geometry.

We note that the theory of algebraically closed fields is not complete, since fields may have different characteristics, so a theory may be decidable without being complete.

Exercises for Section 7.2

Exercise 7.2.1. Show that the theory of successor has an effective procedure for quantifier elimination. This theory has the following axioms:

$(\forall x)S(x) \neq 0$;

$(\forall x)(\forall y)(S(x) = S(y) \to x = y)$;

$(\forall y \neq 0)(\exists x)y = S(x)$;

$(\forall x)S(x) \neq x$;

$(\forall x)S(S(x)) \neq x$;

$\vdots$

$(\forall x)S^n(x) \neq x$;

$\vdots$

7.3. Gödel's Incompleteness Theorems

We now turn to undecidable theories. We will prove Gödel's First Incompleteness Theorem (G1), which states that any effectively axiomatizable theory that contains elementary arithmetic (such as Peano Arithmetic and ZFC) is incomplete and undecidable. Gödel's Second Incompleteness Theorem (G2) states that any effectively axiomatizable theory that contains a sufficient amount of arithmetic cannot prove its own consistency. We will show this holds for Peano Arithmetic.

Just how much arithmetic is sufficient to prove the incompleteness theorems? For G1, it suffices to satisfy the theory of arithmetic in the language $\{+, \times, S, 0, \leq\}$ known as Robinson's Q, given by the following axioms:

(Q_1) $\neg(\exists x)S(x) = 0$;

(Q_2) $(\forall x)(\forall y)(S(x) = S(y) \rightarrow x = y)$;

(Q_3) $(\forall x)(x \neq 0 \rightarrow (\exists y)x = S(y))$;

(Q_4) $(\forall x)(x + 0 = x)$;

(Q_5) $(\forall x)(\forall y)(x + S(y) = S(x + y))$;

(Q_6) $(\forall x)(x \times 0 = 0)$;

(Q_7) $(\forall x)(\forall y)(x \times S(y) = (x \times y) + x)$;

(Q_8) $(\forall x)(\forall y)(x \leq y \leftrightarrow (\exists z)(x + z = y))$.

Note that Q is finitely axiomatizable and hence is effectively axiomatizable. The axioms of Q are exactly what we need to define the operations $+$ and $\times$ and the relation $\leq$. Axioms Q_4 and Q_5 just give the recursive definition of addition and axioms Q_6 and Q_7 give the recursive definition of multiplication. We include the symbol $\leq$ for convenience even though axiom Q_8 may be viewed as a definition of the ordering. Hereafter, the term $S^n 0$ of each $n \in \mathbb{N}$ in PA will be denoted by $\underline{n}$. For simplicity, we will usually abbreviate $\mathbb{N} \models m = n$ as $m = n$.

Peano Arithmetic (PA) adds to Q the following principle:

Definition 7.3.1 (Induction Principle). We will consider the following induction scheme. Informally, for any formula $\varphi(x, y_1, \ldots, y_m)$, and all $y_1, \ldots, y_m$, if

- $\varphi(0, y_1, \ldots, y_m)$, and
- $(\forall x)(\varphi(x, y_1, \ldots, y_m) \to \varphi(S(x), y_1, \ldots, y_m))$,

then we have $(\forall x)\varphi(x, y_1, \ldots, y_m)$. Formally, setting $\vec{y}$ to be the variables $y_1, \ldots, y_m$, we can write this as

$$(\forall \vec{y})((\varphi(0, \vec{y}) \,\wedge\, (\forall x)(\varphi(x, \vec{y}) \to \varphi(S(x), \vec{y}))) \to (\forall x)\varphi(x, \vec{y})).$$

We will assume that PA is consistent and furthermore that $\mathbb{N}$ satisfies the axioms of PA. It then follows by induction on proofs that if $PA \vdash \varphi(\underline{m_0}, \ldots, \underline{m_r})$, then $\mathbb{N} \models \varphi(m_0, \ldots, m_r)$. We will see that, in a certain sense, these assumptions are not provable in PA. Next, we consider how to represent functions and relations in PA.

Example 7.3.2. For any m, n, if $Q \vdash S^m(0) = S^n(0)$ if and only if $m = n$. One direction follows from the assumption that $\mathbb{N} \models PA$. For the other direction, if $m = n$, then $S^m(0)$ is identical with $S^n(0)$, so that $Q \vdash S^m(0) = S^n(0)$ by the equality axiom.

Next we consider the definition of addition. We will show that, for any m, n, and k,

$$m + n = k \iff Q \vdash \underline{m} + \underline{n} = \underline{k}.$$

Example 7.3.3. How do we prove that $2 + 1 = 3$? Axiom Q_4 implies that $S(S(0)) + 0 = S(S(0))$. Then axiom Q_5 implies that $S(S(0)) + S(0) = S(S(S(0)) + 0) = S(S(S(0)))$.

Here is the general result.

Proposition 7.3.4. *For any m, n, and k,*

$$m + n = k \iff Q \vdash \underline{m} + \underline{n} = \underline{k}.$$

Proof. The direction ($\Longleftarrow$) follows from our assumption that $\mathbb{N} \models PA$. The other direction is proved by induction on n.

Base step: For $n = 0$, $m + 0 = m$. Then $\underline{m} + 0 = \underline{m}$ is provable by Axiom Q_5.

Inductive step: Now suppose the result holds for n. We now show that $\underline{m} + \underline{n+1} = \underline{m+n+1}$, that is, $S^m(0) + S^n(0) + 1 = S^{m+n+1}(0)$. By induction, we have

$$Q \vdash S^m(0) + S^n(0) = S^{m+n}(0).$$

Using Axiom Q_5, we obtain that Q implies

$$S^m(0) + S^{n+1}(0) = S^m(0) + S(S^n(0))$$
$$= S(S^m(0) + S^n(0))$$
$$= S(S^{m+n}(0)) = S^{m+n+1}(0). \qquad \square$$

Here is an illustration of the difference between Q and PA.

Example 7.3.5. Addition in the natural numbers is commutative. That is,

$$\mathbb{N} \models (\forall x)(\forall y)(x + y = y + x).$$

This commutative law can be proved in PA using the induction principle, while in Q we can only prove each *instance*. That is, for each m, n,

$$Q \vdash \underline{m} + \underline{n} = \underline{n} + \underline{m}.$$

This fact about Q can be proved by induction on n, since we are assuming that the induction principle is true, although not included in Q. Since the induction principle is included in PA,

$$PA \vdash (\forall x)(\forall y)(x + y = y + x).$$

Other properties of addition provable in PA include the following lemma.

Lemma 7.3.6.

(1) (*Associative Law*) : $(\forall x)(\forall y)(\forall z)(x + (y + z) = (x + y) + z)$.
(2) (*Cancellation Law*) : $(\forall x)(\forall y)(\forall z)((x+y = x+z) \rightarrow y = z)$.
(3) $(\forall x)(\forall y)(x + y = 0 \rightarrow (x = 0 \land y = 0))$.

Proofs are left to the exercises.

Definition 7.3.7. A relation $R \subseteq \mathbb{N}^k$ is *definable in PA* by a formula φ if the following conditions are equivalent for every $u_1, \ldots, u_k \in \mathbb{N}$:

(1) $PA \vdash \varphi(\underline{u_1}, \ldots, \underline{u_k})$;
(2) $\mathbb{N} \models \varphi(\underline{u_1}, \ldots, \underline{u_k})$;
(3) $R(u_1, \ldots, u_k)$.

A function is *definable* in *PA* if its graph is definable in *PA*. A similar definition can be given for definability in Q.

Observe that Proposition 7.3.4 shows that addition is definable in Peano Arithmetic.

To carry out the proof of Gödel's Second Incompleteness Theorem, a theory of arithmetic stronger than Q is necessary (while there is a version of the theorem for Q, it is generally accepted that Q lacks the resources to recognize that the statement asserting its consistency really is its own consistency statement; see, for instance, [2]). We will see that Peano Arithmetic suffices to prove Gödel's Second Theorem. Note that like Q, *PA* is also effectively axiomatizable.

We would like to show the incompleteness and undecidability of arithmetic in the language $\mathcal{L} = \{0, S, +, \times, \leq)$, without exponentiation. We will write $x < y$ for the formula $x \leq y \land x \neq y$.

Thus we need a way to code a sequence $(r_0, r_1, \ldots, r_n)$ of natural numbers in this language, which will be different from the code $0^{r_0} 1 0^{r_1} 1 \cdots 0^{r_n} 1$ used in Chapter 5.

Lemma 7.3.8. *The following relations are both computable and definable in PA and also in Q.*

(1) $x \leq y$;
(2) $\text{rem}(x, y) = z$ *where* z *is the remainder* x *is divided by* y;
(3) $\text{Code}(x, y, z) = \text{rem}(x, 1 + (z + 1)y)$.

Proof.

(1) ($x \leq y$): The defining formula is given by Axiom Q_8. To check this, suppose first that $m \leq n$. Then, for some $k \in \mathbb{N}$, $m + k = n$. It follows from Proposition 7.3.4 that $PA \vdash \underline{m} + \underline{k} = \underline{n}$. Thus $PA \vdash (\exists z)(\underline{m} + z = \underline{n})$.

   To see that the relation is computable, notice that since $x \leq y$ if and only if $x \dot{-} y = 0$, the characteristic function of the relation $x \leq y$ is

$$\chi_{\leq}(x, y) = 1 \dot{-} (x \dot{-} y)$$

(2) ($\text{rem}(x, y)$): The formula that defines this relation in PA is

$$\text{rem}(x, y) = z \text{ if and only if } (\exists q)((x = yq + z) \wedge (0 \leq z < y)).$$

   We have already shown in the previous chapter that $\text{rem}(x, y)$ is computable.

(3) ($\text{Code}(x, y, z)$): The definition of *Code* in terms of *rem* is given above, and so it is clear that *Code* is definable in *PA*. Since addition, multiplication and *rem* are all computable, so is *Code*. □

To code a finite sequence of numbers, we will need the following Theorem.

Theorem 7.3.9 (Chinese Remainder Theorem). *Let* m_1, $\ldots, m_n$ *be pairwise relatively prime (i.e.,* $\gcd(m_i, m_j) = 1$ *for*

$i \neq j$). *Then for any* $a_1, \ldots, a_n \in \mathbb{N}$, *the system of equations*

$$x \equiv a_1 \pmod{m_1}$$

$$\vdots$$

$$x \equiv a_n \pmod{m_n}$$

has a unique solution modulo $M = m_1 \cdot \ldots \cdot m_n$.

See [11, p. 248] for a proof of this result.

Theorem 7.3.10. *For any sequence* $k_1, k_2, \ldots, k_n$ *of natural numbers, there exist natural numbers* a, b *such that* $\text{Code}(a, b, 0) = n$ *and* $\text{Code}(a, b, i) = k_i$ *for* $i = 1, 2, \ldots, n$.

Proof. We prove this using the Chinese Remainder Theorem, as follows: Let $s = \max\{n, k_1, \ldots, k_n\}$ and set $b = s!$. It is not hard to see that the numbers

$$s! + 1, 2s! + 1, \ldots, (n + 1)s! + 1$$

are pairwise relatively prime. Then by the Chinese Remainder Theorem, there is a unique solution x to

$$x \equiv n \pmod{s! + 1}$$
$$x \equiv k_1 \pmod{2s! + 1}$$
$$\vdots$$
$$x \equiv k_n \pmod{(n + 1)s! + 1}$$

modulo $M = \prod_{i=1}^{n+1} is! + 1$. Let $a = x$.

We now check that a and b are the desired values. First,

$$\text{Code}(a, b, 0) = \text{rem}(a, b + 1)$$

$$= \text{rem}(a, s! + 1) = n.$$

For $i = 1, \ldots, n$,

$$\text{Code}(a, b, i) = \text{rem}(a, (i + 1)b + 1)$$

$$= \text{rem}(a, (i + 1)s! + 1) = k_i. \qquad \square$$

We will make use of the following in the proof of Gödel's First Incompleteness Theorem.

Theorem 7.3.11.

(1) *The computably enumerable relations are exactly the relations definable in PA (and also in Q).*
(2) *The computable functions are exactly the functions definable in PA (and also in Q).*

Proof. We prove these two statements in tandem. We first show that the right-to-left containment of both statements.

($\supseteq$): Suppose that $R \subseteq \mathbb{N}^k$ is definable in *PA*. Then there is some $\mathcal{L}$-formula φ_f such that for all $u_1, \ldots, u_k \in \mathbb{N}$,

$$PA \vdash \varphi(\underline{u_1}, \ldots, \underline{u_k}) \iff R(u_1, \ldots, u_k).$$

Then R is computably enumerable by Proposition 7.2.3.

Next, if the partial function f is definable in *PA*, then the graph of f is thus computably enumerable by what we have just shown above, and therefore f is computable by Proposition 5.4.4.

($\subseteq$): For the other direction, we now handle the two statements in the opposite order. We first show that all partial computable functions are definable in *PA* by induction. It is not hard to show that the initial functions are definable.

(1) (Constant Function): The defining formula $\varphi_c(x, y)$ for the constant function $c(x) = 0$ is

$$(y = \underline{0}).$$

(2) (Projective Function): The defining formula $\varphi_{p_j^k}(x_1, \ldots, x_k, y)$ for the projection function $p_j^k(x_1, \ldots, x_k) = x_j$ is

$$(y = x_j).$$

(3) (Successor Function): The defining formula $\varphi_S(x, y)$ for the successor function $S(x) = x + 1$ is $y = S(x)$.

For the induction step of the proof, we must show that the set of *PA*-definable functions is closed under the production rules for the set of partial computable functions.

(4) (Composition): Suppose that f and $g_1, g_2, \ldots, g_k$ are definable in *PA* by the formulas φ_f and $\varphi_{g_1}, \ldots, \varphi_{g_k}$, respectively and that h is the function $h(\vec{x}) = f(g_1(\vec{x}), \ldots, g_k(\vec{x}))$. Then the defining formula $\varphi_h(\vec{x}, y, z)$ for h is

$$(\exists y_1)(\exists y_2) \cdots (\exists y_k)(\varphi_{g_1}(\vec{x}, y_1) \wedge \cdots \wedge \varphi_{g_k}(\vec{x}, y_k)$$
$$\wedge \, \varphi_f(y_1, \ldots, y_k, z)).$$

(5) (Primitive Recursion): Suppose that f and g are definable in *PA* by φ_f and φ_g, respectively, and that h is the function defined by recursion with

$$h(\vec{x}, 0) = f(\vec{x}),$$
$$h(\vec{x}, y + 1) = g(\vec{x}, y, h(\vec{x}, y)).$$

To define h, we will use a pair of numbers a and b that code up the sequence

$$h(\vec{x}, 0), h(\vec{x}, 1), \ldots, h(\vec{x}, y),$$

via the function *Code*, where $h(\vec{x}, 0) = f(\vec{x})$ and $h(\vec{x}, n + 1) = g(\vec{x}, n, h(\vec{x}, n))$ for every $n < y$. Thus the defining formula $\varphi_h(\vec{x}, y, z)$ for h is

$(\exists a)(\exists b)(\varphi_f(\vec{x}, \text{Code}(a, b, 1))$
$\qquad \wedge \, (\forall i < y)(\varphi_g(\vec{x}, i, \text{Code}(a, b, i + 1), \text{Code}(a, b, i + 2))$
$\qquad \wedge \, (z = \text{Code}(a, b, y + 1))).$

(6) (Unbounded Search): Left to the reader.

We have now provided a definition φ_f for each partial computable function f. It remains to check that in each case, $f(m_1, \ldots, m_k) = n$ if and only if $PA \vdash \varphi_f(\underline{m_1}, \ldots, \underline{m_k}, \underline{n})$. The direction ($\Longleftarrow$) follows as usual because $\mathbb{N} \models PA$.

The other direction is shown by induction on the computable functions. For Case 1, $0 = 0$ is provable by the Equality Axiom. Case 2 is similar. Cases 3 and 6 are left as exercises. Case 5 follows from Lemma 7.3.8. Here is the proof of a simple version of Case 4.

Suppose that $h(x) = f(g(x))$ and that, by induction, $g(x) = y$ is definable in PA by the formula $\varphi_g(x, y)$ and $f(y) = z$ is definable by the formula $\varphi_f(y, z)$. Then, we are given the defining formula $\varphi_h(x, z)$ as $(\exists y)(\varphi_g(x, y) \wedge \varphi_f(y, z))$.

Now suppose that $g(m) = k$ and $f(k) = n$, so that $h(m) = n$. By induction, PA proves both $\varphi_g(\underline{m}, \underline{k})$ and $\varphi_f(\underline{k}, \underline{n})$. It follows directly that PA proves $\varphi_h(\underline{m}, \underline{n})$.

This completes the proof that the class of functions definable in PA includes all the initial functions and is closed under the production rules. Therefore every partial computable function is definable in PA.

Finally, if R is a computably enumerable relation, then, by Theorem 5.4.3, there is a computable function f such that $R = \text{dom}(f)$. As f is definable in PA by what we have shown above, it follows that R is definable in PA. $\qquad\square$

Using this result, we can now obtain basic versions of Gödel's First Incompleteness Theorem as well as Church's Undecidability Theorem.

Theorem 7.3.12 (Church's Theorem). *If PA is consistent, then it is not decidable.*

Proof. Let $\text{Th}(PA) = \{\varphi : PA \vdash \varphi\}$. Let H be the c.e. noncomputable halting set. Since H is c.e., by part (1) of Theorem 7.3.11, there is a formula θ of arithmetic such that,

for any e,
$$e \in H \iff PA \vdash \theta(\underline{e}),$$
that is,
$$e \in H \iff \theta(\underline{e}) \in \mathrm{Th}(PA).$$

Since H is not computable, it follows that $\mathrm{Th}(PA)$ is not computable. □

The above result in fact holds for Q in the place of PA. Using this observation, we can prove the following corollary.

Corollary 7.3.13. *Predicate logic is undecidable.*

Proof. As above, since H is c.e., the relation $e \in H$ is definable in Q by part 1 of Theorem 7.3.11. Thus we have

$$e \in H \iff Q \vdash \theta(\underline{e})$$

for some formula θ of arithmetic. Since Q is finite, we can let φ be the conjunction of the axioms of Q. Then we have

$$e \in H \iff \vdash (\varphi \to \theta(\underline{e})).$$

Thus, the decision problem for predicate logic is unsolvable. □

We can now give our first version of the First Incompleteness Theorem.

Theorem 7.3.14 (Gödel's First Incompleteness Theorem, Version 1). *If PA is consistent, then it is not complete.*

Proof. By Theorem 7.3.12, PA is not decidable. It then follows from Theorem 7.2.4 that PA cannot be complete. □

If we let φ be a sentence of arithmetic such that PA cannot prove either φ or $\neg\varphi$, then without loss of generality, $\mathbb{N} \models \varphi$, so that we have a sentence which is true but not provable. There is in fact a *canonical* sentence not provable in PA, namely a sentence which asserts the consistency of PA itself. This is the so-called Second Incompleteness Theorem of Gödel. We will examine a direct proof of the Incompleteness Theorem which will lead us to this result. This direct proof is based on a diagonal argument. As a warm-up, let us first consider a closely related diagonal argument. Let $\varphi_e(x)$ be the formula φ with one free variable x such that $\langle \varphi \rangle = e$; if the formula coded by e does not have exactly one free variable, then $\varphi_e(x)$ is false for every x.

Proposition 7.3.15. *There is no formula ψ such that, for all $e, n \in \mathbb{N}$, $\mathbb{N} \models \psi(e, n) \leftrightarrow \varphi_e(n)$.*

Proof. Suppose by way of contradiction that there were such a formula ψ and let $\varphi(x)$ be the formula $\neg\psi(x, x)$. Then φ is the formula φ_e for some e, so that, for all $n \in \mathbb{N}$,

$$\varphi_e(n) \iff \neg\psi(n, n).$$

Setting $n = e$, we have

$$\psi(e, e) \iff \varphi_e(e) \iff \neg\psi(e, e). \qquad \square$$

Theorem 7.3.16 (Gödel's First Incompleteness Theorem, Version 2). *If PA is consistent, then there is a sentence ψ_G true in the standard model of arithmetic such that ψ_G is equivalent to its own unprovability, and hence not provable in PA.*

Proof. We modify the diagonal argument in Proposition 7.3.15 above as follows to obtain a sentence ψ_G, called the *Gödel sentence*, which is equivalent to its own unprovability. First observe that the relation "$PA \vdash \varphi_e(n)$" is computably enumerable by Proposition 7.2.3. Thus by Theorem 7.3.11, there is a formula $\psi(e, n)$ such that, for all e and n,

$$\mathbb{N} \models \psi(e, n) \iff PA \vdash \varphi_e(\underline{n}).$$

As in the proof of Proposition 7.3.15 above, consider the formula φ so that

$$\varphi(n) \iff \neg\psi(n, n).$$

Then as in Proposition 7.3.15, φ is the formula φ_e for some e, so that, for all $n \in \mathbb{N}$,

$$\varphi_e(n) \iff \neg\psi(n, n).$$

Setting $n = e$ we have

$$\varphi_e(e) \iff \neg\psi(e, e) \iff PA \nvdash \varphi_e(\underline{e}).$$

So $\varphi_e(e)$ is the desired sentence ψ_G, a sentence which is equivalent to its own unprovability. This can also be written as

$$\mathbb{N} \models \neg\psi_G \iff PA \vdash \psi_G.$$

We claim that neither ψ_G not $\neg\psi_G$ is provable in PA.

Suppose first that $PA \vdash \psi_G$. Then by the definition of ψ_G, $\mathbb{N} \models \neg\psi_G$. However, since $\mathbb{N} \models PA$, it also follows that $\mathbb{N} \models \psi_G$, a contradiction. Next suppose that $PA \vdash \neg\psi_G$. Then $\mathbb{N} \models \neg\psi_G$, but this implies that $PA \vdash \psi_G$, which contradicts the consistency of PA. $\qquad\square$

We now turn to Gödel's Second Incompleteness Theorem. Informally, the theorem states that if PA is consistent, then it cannot prove that it is consistent. To express this formally, we need to formalize that statement that PA is consistent *within* PA. Recall the general formula $\psi(e, n)$ such that $\mathbb{N} \models \psi(e, n)$ if and only if $PA \vdash \phi_e(\underline{n})$. When a sentence ψ asserts in this fashion that there is a proof of another sentence φ, we say that there is a proof of a sentence φ within PA, when $PA \vdash \psi(e, n)$.

Let $\theta(x)$ be the statement $x = S(0)$, so that $\theta(x)$ is $\phi_i(x)$ for some i. Of course $PA \vdash \neg(0 = S0)$ by Axiom Q_1, so that PA is consistent if and only if $PA \nvdash \phi_i(0)$ (any contradiction will do here, but this one is standard). Then by definition of the formula ψ above, we have

$$\mathbb{N} \models \psi(i, 0) \iff PA \vdash \varphi_i(0),$$

and hence PA is consistent if and only if $\mathbb{N} \models \neg\psi(i, 0)$. Thus, we will write $\neg\psi(i, 0)$ as $\mathrm{Con}(PA)$, the expression of the consistency of PA within arithmetic.

We will make use of the following lemma, which states that the proof of Gödel's First Incompleteness Theorem can be shown *within PA*. A majority of the work in proving Gödel's Second Incompleteness Theorem is in the proof of this lemma, which is rather technical; see, for instance, Section 5.3 of [18] for details.

Lemma 7.3.17. $PA \vdash \mathrm{Con}(PA) \to \psi_G$.

We are now ready to state and prove Gödel's Second Incompleteness Theorem.

Theorem 7.3.18 (Gödel's Second Incompleteness Theorem). *If PA is consistent, then PA does not prove its own consistency.*

Proof. Suppose that $PA \vdash \mathrm{Con}(PA)$. Then by Lemma 7.3.17, $PA \vdash \psi_G$, which, under the assumption that PA is consistent, contradicts Gödel's First Incompleteness Theorem. It thus follows that $PA \nvdash \mathrm{Con}(PA)$. □

Exercises for Section 7.3

Exercise 7.3.1. Show that, for any m, n, and k,

$$m \times n = k \iff Q \vdash \underline{m} \times \underline{n} = \underline{k}.$$

Exercise 7.3.2. (Associative Law): $(\forall x)(\forall y)(\forall z)x + (y + z) = (x + y) + z$.

Exercise 7.3.3. (Cancellation Law): $(\forall x)(\forall y)(\forall z)[(x + y = x + z \to y = z]$.

Exercise 7.3.4. Prove in Peano Arithmetic that $(\forall x)(\forall y)[x + y = 0 \to (x = 0 \land y = 0)]$.

Exercise 7.3.5. Give the details to show that the formula for the remainder given in Lemma 7.3.8 works.

Exercise 7.3.6. Give the details for Case (6) in the proof of Theorem 7.3.11.

Exercise 7.3.7. Show that the formulas in Cases (3) and (4) from Theorem 7.3.11 are provable, that is (in Case (3), show that if $S(x) = y$, then this can be proved in PA.

Chapter 8

Algorithmic Randomness

In this chapter, we will briefly the discuss the basics of the theory of algorithmically random strings. Although this is not a topic typically covered in an introduction to the foundations of mathematics, the theory of algorithmic randomness provides interesting examples of the phenomena of undecidability and incompleteness that we considered in the previous two chapters.

How exactly do we define a finite binary string to be random? One possibility is to take such a string to be random if it is produced by a paradigmatically random process such as the tosses of a fair coin. For some purposes, such a definition is sufficient. For our purposes, however, we would like a definition that picks out a fixed collection of binary strings as the random ones (although this is not quite what the theory of algorithmically random finite strings delivers for us).

As a first step, consider the following binary strings of length 50:

(1) 00000 00000 00000 00000 00000 00000 00000 00000 00000 00000
(2) 01010 10101 01010 10101 01010 10101 01010 10101 01010 10101
(3) 10100 00011 01010 00110 10110 10001 11110 00011 11101 00011
(4) 00100 10000 11111 10110 10101 00010 00100 00101 10100 01100
(5) 01001 00101 10101 11111 11010 10100 11110 01111 11111 10010

Although it is possible to obtain each of these strings by tossing an unbiased coin (where we take 0 to stand for heads and 1 to stand for tails), it is *highly* unlikely that we would produce (1) and (2) in this way. In fact, (1) and (2) just do not appear to be the sort of strings produced by the tosses of an unbiased coin. It is less clear whether (3)–(5) can be thus produced. For instance, (3) has the same number of 0s and 1s, and since most strings of length 50 have close to the same number of 0s and 1s, we might surmise that (3) was produced by tossing a fair coin. However, (3) encodes the parity of the number of words of each of the 50 states in alphabetical order (e.g., "Alabama" has an odd number of letters and thus corresponds to a "0"). Although the digits of (4) are less evenly distributed than those of (3) (29 0s and 21 1s), this is not inconsistent with having a random origin. However, once we learn that (4) consists of the first 50 bits of the binary expansion of π, we would still be surprised if we were to produce this string by coin tossing. Lastly, (5) is slightly more biased than (3) or (4), but this was (purportedly) produced by encoding some atmospheric noise as a binary string (at least according to random.org).

What is the upshot of this discussion? One is that strings that appear to be random such as (3) and (4) are no longer thought to be random when they can be given some simpler description. This is one key idea behind the definition of an algorithmically random string: random strings are difficult to describe. Here we must exercise some caution: we consider a description of a string to be one that allows us to reconstruct the string unambiguously, like a blueprint for reproducing the string. The other key idea behind the definition of an algorithmically random string is that the descriptions we consider are computational, given in some way by a Turing machine.

For a detailed study of Kolmogorov complexity, see the book of Li, Ming and Vitanyi [23]. Algorithmic randomness and its connection with computability are developed by Downey and Hirschfeldt [10] and also Nies [26].

8.1. Kolmogorov Complexity

We now make the above ideas precise by defining the Kolmogorov complexity of string, which allows us to define the complexity of effectively reproducing that string. We begin with an example.

Example 8.1.1. We define a Turing machine $M : \{0,1\}^* \to \{0,1\}^*$ that gives short descriptions of strings of the form 1^n for $n \in \mathbb{N}$. Given an input string $\sigma \in \{0,1\}^*$, M interprets this as some $n \in \mathbb{N}$ written in (reverse) binary and outputs n 1s. Note that if $n = 2^k$ for some k, then the input string representing n will have length $k+1$ and will yield as output a string of length 2^k, a considerable gain. Thus, we can think of the input string as a compressed version of the output string.

The previous example suggests the following definition.

Definition 8.1.2. Let $M : \{0,1\}^* \to \{0,1\}^*$ be a Turing machine and let $\sigma \in \{0,1\}^*$. We define the *Kolmogorov complexity of σ relative to M* to be

$$C_M(\sigma) = \min\{|\tau| : M(\tau){\downarrow} = \sigma\},$$

where $|\tau|$ is the length of τ. Moreover, if $\sigma \notin \mathrm{ran}(M)$, then we set $C_M(\sigma) = \infty$.

If $M(\tau) = \sigma$, we can think of τ as an M-description of σ. Then $C_M(\sigma)$ is the length of the shortest M-description of σ.

Example 8.1.3. If, as in the previous example, $M(\sigma_n) = 1^n$, where σ_n is the binary representation of $n \in \mathbb{N}$, then

$$C_M(1^n) = \lfloor \log_2(n) \rfloor + 1$$

(and $C_M(\sigma) = \infty$ if $\sigma \neq 1^n$ for every $n \in \mathbb{N}$).

One might reasonably worry here that this measure of complexity is completely dependent on our choice of underlying machine. This worry is not unjustified. However, this worry can be mitigated if we define complexity in terms of a universal Turing machine.

Let $(M_e)_{e \in \mathbb{N}}$ be a computable enumeration of all Turing machines (effectively coded in some way). Then we define $U : \{0,1\}^* \to \{0,1\}^*$ by

$$U(1^e 0 \tau) \simeq M_e(\tau)$$

for every $e \in \mathbb{N}$ and $\sigma \in \{0,1\}^*$.

Definition 8.1.4. The *Kolmogorov complexity* of σ is defined to be

$$C(\sigma) := C_U(\sigma) = \min\{|\tau| : U(\tau){\downarrow} = \sigma\}.$$

To what extent does this definition address the concern with machine dependence? The following theorem, first proved by Kolmogorov in [22], provides an answer.

Theorem 8.1.5 (Invariance Theorem). *For any Turing machine $M : \{0,1\}^* \to \{0,1\}^*$, there is some $c \in \mathbb{N}$ such that*

$$C(\sigma) \leq C_M(\sigma) + c$$

for every $\sigma \in \{0,1\}^*$.

Proof. Given $M : \{0,1\}^* \to \{0,1\}^*$, let e be such that $M_e = M$. Then for all $\tau \in \{0,1\}^*$, $U(1^e 0 \tau) \simeq M_e(\tau)$. Given $\sigma \in \{0,1\}^*$, let σ^* be the shortest string such that $M_e(\sigma^*) = \sigma$. Then $C_M(\sigma) = |\sigma^*|$. Since $U(1^e 0 \sigma^*) = M_e(\sigma^*) = \sigma$, it follows that

$$C(\sigma) \leq |1^e 0 \sigma^*| = |\sigma^*| + e + 1 = C_M(\sigma) + e + 1.$$

The conclusion follows by setting $c = e + 1$. $\square$

We now find a simple upper bound for C.

Proposition 8.1.6. *There is some $c \in \mathbb{N}$ such that*

$$C(\sigma) \leq |\sigma| + c$$

for every $\sigma \in \{0,1\}^*$.

Proof. Let $M_{id} : \{0,1\}^* \to \{0,1\}^*$ be the identity machine, which satisfies $M(\sigma) = \sigma$ for every $\sigma \in \{0,1\}^*$. Then by the Invariance Theorem, there is some $c \in \mathbb{N}$ such that

$$C(\sigma) \leq C_{M_{id}}(\sigma) + c = |\sigma| + c. \qquad \square$$

8.2. Incompressible Strings

We now formalize the notion of an incompressible string, which will serve as the basis of our definition of random strings.

Definition 8.2.1. Let $c \in \mathbb{N}$. We say that $\sigma \in \{0,1\}^*$ is *c-incompressible* if

$$C(\sigma) \geq |\sigma| - c.$$

Do c-incompressible strings exist? Let us consider the case that $c = 0$. Note that $\sigma \in \{0,1\}^*$ is 0-incompressible if for every $\tau \in \{0,1\}^*$ such that $U(\tau){\downarrow} = \sigma$, it follows that $|\tau| \geq |\sigma|$. We claim that there is at least one 0-incompressible string of each length. Let us consider strings of length n. There are

$$1 + 2 + \cdots + 2^{n-1} = 2^n - 1$$

strings of length $< n$. Hence, in the worst case, if each string of length $< n$ is mapped by U to a unique string of length n, then one string of length n remains and is hence 0-incompressible. We now generalize this line of reasoning to prove the following proposition.

Proposition 8.2.2. *For each $c \in \mathbb{N}$ and $n \in \mathbb{N}$, there are at least $2^n - (2^{n-c} - 1)$ c-incompressible strings of length n.*

Proof. There are

$$1 + 2 + \cdots + 2^{(n-c)-1} = 2^{n-c} - 1$$

strings of length $< n - c$. Thus the number of strings of length n to which U maps a string of length $< n - c$ is at most $2^{n-c} - 1$.

Hence, the number of strings of length n to which U maps no string of length $< n - c$ is at least $2^n - (2^{n-c} - 1)$. Since all such strings are c-incompressible, the conclusion follows.   □

Note that $2^n - (2^{n-c} - 1) = 2^n(1 - 2^{-c}) + 1 > 2^n(1 - 2^{-c})$. From this we can conclude that for each $n \in \mathbb{N}$,

- at least $1/2$ of the strings of length n are 1-incompressible;
- at least $3/4$ of the strings of length n are 2-incompressible;
- at least $7/8$ of the strings of length n are 3-incompressible;

and so on. Thus, if we consider, say, strings of length 100, at least $1023/1024$ of them have complexity greater than 90. It is thus quite easy to produce such a string: toss a fair coin 100 times, and an overwhelming majority of the time, a 10-incompressible string will be produced.

Hereafter, we will identify the random strings with sufficiently incompressible strings. Just how incompressible? We cannot give a precise answer, but for each c, for sufficiently long c-incompressible strings behave, for all practical purposes, like statistically random strings.

Given that it is easy to randomly produce an incompressible string, we might also ask whether we can produce incompressible strings algorithmically. Let $\mathcal{S}_c$ be the collection of c-incompressible strings. Is there a computable function $f : \mathbb{N} \to \{0,1\}^*$ such that $\mathrm{ran}(f) \subseteq \mathcal{S}_c$?

Definition 8.2.3. A set $S \subseteq \{0,1\}^*$ is *immune* if it is infinite and has no infinite computably enumerable subset.

We will show the following, which immediately implies a negative answer to the above question.

Theorem 8.2.4. *For each $c \in \mathbb{N}$, the collection of c-incompressible strings is immune.*

Proof. Fix $c \in \mathbb{N}$. First note that $\mathcal{S}_c$ is infinite, since by Proposition 8.2.2, there is a c-incompressible string of every length.

Now suppose that $\mathcal{S}_c$ contains an infinite computably enumerable subset $\{\tau_0, \tau_1, \ldots\}$. That is, there is some (total) computable function $f : \mathbb{N} \to \{0,1\}^*$ such that $f(i) = \tau_i$ for every i. Without loss of generality, we can assume that $|\tau_i| < |\tau_{i+1}|$ for every i. Indeed, we can define a function $\hat{f} : \mathbb{N} \to \{0,1\}^*$ that copies f as long as f enumerates a string with length longer than the lengths of the previously enumerated strings. That is, if f enumerates a string of length less than or equal to a previously enumerated string, $\hat{f}$ will not enumerate this string but will wait for a longer string to be enumerated by f. Since $\mathrm{ran}(\hat{f}) \subseteq \mathrm{ran}(f) \subseteq \mathcal{S}_c$, we can assume that f has this property (replacing f with $\hat{f}$ if it does not). Note that $|\tau_i| < |\tau_{i+1}|$ for every i implies that $|\tau_i| \geq i$ for every i.

Now, using the function f, we can define a Turing machine M_f that maps n (written in binary) to $f(n)$ for each n. Then

$$C_{M_f}(\tau_n) \leq \lfloor \log_2(n) \rfloor + 1$$

and hence

$$C(\tau_n) \leq \lfloor \log_2(n) \rfloor + d \tag{7.1}$$

for some $d \in \mathbb{N}$. However, since each τ_n is c-incompressible,

$$C(\tau_n) \geq |\tau_n| - c \geq n - c, \tag{7.2}$$

where the latter inequality follows from the observation at the end of the previous paragraph. But for sufficiently large n,

$$n - c > \lfloor \log_2(n) \rfloor + d,$$

which, together with (7.1) and (7.2), yields a contradiction. Thus, $\mathcal{S}_c$ contains no infinite computably enumerable subset. $\square$

Corollary 8.2.5. *There is no total computable $f : \mathbb{N} \to \{0,1\}^*$ whose range consists only of c-incompressible strings.*

Corollary 8.2.6. *The function $C : \{0,1\}^* \to \mathbb{N}$ is not computable.*

Proof. If we could compute the value $C(\sigma)$ for each $\sigma \in \{0,1\}^*$, we could immediately determine which strings are c-incompressible (for any fixed c). Then we could define a computable enumeration of *all* c-incompressible strings, which is impossible. □

One other consequence of Theorem 8.2.4 is that we have a second example of a set that is computably enumerable but not computable (the first being the Halting Problem).

Theorem 8.2.7. *For each $c \in \mathbb{N}$, the collection of c-compressible strings is computably enumerable but not computable.*

Proof. Recall that a set $S \subseteq \{0,1\}^*$ is computable if and only if S and $\{0,1\}^* \setminus S$ are both computably enumerable. Since the collection of c-incompressible strings is immune, it cannot be computably enumerable. It follows that the collection of c-compressible strings is not computable.

To show that this collection is computably enumerable, observe that $\sigma \in \{0,1\}^*$ is c-compressible if and only if

$$(\exists \tau)(\exists s)(U(\tau)\!\downarrow \text{ in } < s \text{ steps } \wedge \ U(\tau) = \sigma \ \wedge \ |\tau| < |\sigma| - c).$$

Note that the relation $R(\sigma, \tau, s)$ given by

$$U(\tau)\!\downarrow \text{ in } < s \text{ steps } \wedge \ U(\tau) = \sigma \ \wedge \ |\tau| < |\sigma| - c$$

is a computable relation. Hence, by Exercise 5.4.1 and Theorem 5.4.3, the collection of c-compressible strings is computably enumerable. □

8.3. Kolmogorov Complexity and Incompleteness

In this final section, we will prove yet another incompleteness theorem due to Gregory Chaitin [5] known as Chaitin's Incompleteness Theorem (CIT), which provides us with infinitely many undecidable sentences in the language of arithmetic.

In order to prove CIT, we need to express statements about the Kolmogorov complexity of strings in Peano arithmetic.

By Theorem 7.3.11 there is some formula $\psi(x, y)$ (of the form $(\exists z)\theta(x, y, z)$, where θ has no unbounded quantifiers) such that

$$U(\sigma) = \tau \iff PA \vdash \psi(< \sigma >, < \tau >).$$

Note that $C(\sigma) \leq n$ if and only if there is some τ such that $|\tau| = n$ and $U(\tau) = \sigma$. Thus, using our coding of members of $\{0, 1\}^*$ and the representations of $|\cdot|$ and U, we can express this by a formula in the language of arithmetic; for ease of presentation, we will simply write this formula as $C(\sigma) \leq n$.

Theorem 8.3.1 (Chaitin's Incompleteness Theorem). *If PA is consistent, then there is some $L \in \mathbb{N}$ such that*

$$PA \nvdash C(\sigma) \geq L$$

for any $\sigma \in \{0, 1\}^*$ and hence for all $n \geq L$,

$$PA \nvdash C(\sigma) \geq n$$

for any $\sigma \in \{0, 1\}^*$.

We first prove a lemma.

Lemma 8.3.2. *If PA is consistent, then for $n \in \mathbb{N}$ and $\sigma \in \{0, 1\}^*$,*

$$PA \vdash C(\sigma) \geq n \;\Rightarrow\; \mathbb{N} \models C(\sigma) \geq n.$$

That is, if PA proves that the complexity of σ is at least n, then it is really the case that the complexity of σ is at least n.

Proof. Suppose that $PA \vdash C(\sigma) \geq n$ but in fact we have that $\mathbb{N} \models C(\sigma) < n$. The sentence expressing that $C(\sigma) < n$ has the form $(\exists x)\theta(x)$. However, by Definition 7.3.7 and Theorem 7.3.11, for any such sentence, we have

$$\mathbb{N} \models (\exists x)\theta(x) \Rightarrow PA \vdash (\exists x)\theta(x).$$

Hence $PA \vdash C(\sigma) < n$, but this contradicts our assumption that PA is consistent. $\square$

Proof of CIT. Suppose that for every $n \in \mathbb{N}$, there is some $\sigma \in \{0, 1\}^*$ such that

$$PA \vdash C(\sigma) \geq n.$$

We define a machine $M : \{0, 1\}^* \to \{0, 1\}^*$ that on input τ enumerates theorems of PA until it finds a proof of $C(\sigma) \geq k$ for some σ and some $k > 2|\tau|$. Then M outputs σ.

If $M(\tau)\!\downarrow = \sigma$, then $PA \vdash C(\sigma) \geq k$ for some $k > 2|\tau|$, and hence by by Lemma 8.3.2, it follows that $C(\sigma) \geq k > 2|\tau|$. By the Invariance Theorem, there is some $d \in \mathbb{N}$ such that

$$C(\sigma) \leq C_M(\sigma) + d$$

for every $\sigma \in \{0, 1\}^*$. Now, if we give M some input δ with $|\delta| = d$, we are guaranteed by our initial assumption to find some $\sigma \in \{0, 1\}^*$ such that $C(\sigma) > 2|\delta| = 2d$. However, since $M(\delta)\!\downarrow = \sigma$ we also have

$$C(\sigma) \leq C_M(\sigma) + d \leq |\delta| + d = 2d,$$

which is impossible. Thus it must be the case that PA can only prove $C(\sigma) \geq n$ for finitely many n, bounded by some $L \in \mathbb{N}$.　　　　　　　　　　　　　　□

Chapter 9

Nonstandard Numbers

In this chapter, we explore the structure of nonstandard models of $\mathbb{N}$ and $\mathbb{R}$. Such structures exist by the Compactness Theorem. The primary focus here will be on the construction of nonstandard models using ultraproducts. Nonstandard models of the reals include so-called *infinitesimal* elements which provide an alternative approach to the study of calculus.

In the first section, we examine nonstandard models of the natural numbers in certain limited languages, such as $\{0, S\}$, $\{0, S, <\}$ and $\{0, S, +, <\}$. Here we look at models satisfying some of the basic axioms from Robinson Arithmetic together with ordering axioms. We also look at models of Peano Arithmetic. Finally, we use ultraproducts to give a concrete presentation of a nonstandard model $\mathbb{N}^*$ of the natural numbers. The nonstandard model $\mathbb{N}^*$ may be viewed as an extension of the standard model $\mathbb{N}$ obtained by adding *infinite* elements.

In the second section, we again use ultraproducts to present a nonstandard model $\mathbb{R}^*$ of the reals with added function symbols. This leads to a brief introduction to nonstandard analysis. The nonstandard model of the reals may be viewed as an extension of the standard model $\mathbb{R}$ which contains infinite as well as *infinitesimal* numbers.

9.1.   Nonstandard Natural Numbers

In this section, we specify certain limited sets of axioms for the
natural numbers and examine the nonstandard models which
can satisfy those axioms. In particular, we are interested in finite
sets of axioms. We are also interested in the Induction Principle.
So we do not assume that $\mathbb{N}$ satisfies the Induction Principle
unless otherwise indicated.

The first part of the language of natural numbers is the
language of successor: $\{0, S\}$. Recall the first three axioms of
Robinson arithmetic:

$(Q_1)\ \neg(\exists x)S(x) = 0;$
$(Q_2)\ (\forall x)(\forall y)(S(x) = S(y) \rightarrow x = y);$
$(Q_3)\ (\forall x)(x \neq 0 \rightarrow (\exists y)x = S(y)).$

The language of successor includes names $\underline{n} = S^n(0)$ for the
natural numbers. In any language with addition, there is an
alternative definition of n as

$$\underbrace{1 + \cdots + 1}_{n \text{ times}}.$$

Any model $\mathcal{M}$ of the three axioms above must include a copy

$$\{0, S(0), S(S(0)), \ldots\}.$$

of the natural numbers. Axiom (Q_2) ensures that these are all
distinct.

Recall from Chapter 3 that the Compactness Theorem
implies that there is a model of these axioms with a nonstan-
dard element. Let ψ_n be the sentence $c \neq \underline{n}$. By the Compactness Theorem, $\mathrm{Th}(\mathbb{N}) \cup \{\psi_n : n \in \mathbb{N}\}$ is consistent. So there is a
nonstandard model $\mathcal{M}$ with an infinite element.

What does a nonstandard model of successor look like? If c is
a nonstandard element, then $c \neq 0$, so c has a predecessor $S^{-1}(c)$.
Then $S^{-1}(c)$ is still infinite so it has a predecessor $S^{-2}(c)$, and
so on.

Definition 9.1.1. Let f be an injective function on a set M. For any element $x \in M$, the *orbit* $\mathcal{O}_f(x)$ is

$$\mathcal{O}_f(x) = \{y : (\exists n)(y = S^n(x) \ \lor \ x = S^n(y))\}.$$

For models of successor, we will just write $\mathcal{O}_S(x)$ as $\mathcal{O}(x)$. For example, in a nonstandard model of the natural numbers, the orbit of any standard natural number n is $\mathbb{N}$.

What do the other orbits look like? Axiom (Q_2) states that S is an injection. In general, orbits can have the following potential forms:

(a) A copy of ω, $\{c, S(c), S^2(c), \dots\}$ where c has no predecessor; we will call this an orbit of type ω, or just an ω-orbit.

(b) A copy of $\mathbb{Z}$, $\{\dots, S^{-1}(c), c, S(c), \dots\}$; we will call this an orbit of type $\mathbb{Z}$, or just a $\mathbb{Z}$-orbit.

(c) A finite cycle $\{c, S(c), \dots, S^{n-1}(c)\}$ where $S^n(c) = c$; we will call this an orbit of type n, or, in general, a finite orbit.

By Axiom (Q_1), the orbit of $\{0\}$ has type ω. By Axiom (Q_3), the orbit of any infinite element is either finite or has type $\mathbb{Z}$.

Finite orbits are not natural. One way to avoid them is by adding infinitely many new axioms of the form $(\forall x)S^{n+1}(x) \neq x$. A second approach is to include the Induction Principle. As we have often proved things in this book using induction, we accept it as given. This means that we can use the Induction Principle to prove properties of the *standard* natural numbers. Thus we showed in Example 6.3.2 that $S^m(0) = S^n(0)$ implies $m = n$. Recall that a standard natural number n has a name $\underline{n} = S^n(0)$ in the language of successor. Without assuming that the Induction Principle holds in a nonstandard model $\mathcal{M}$, we can still show that for any n and any k,

$$S^{n+1}(\underline{k}) \neq \underline{k}.$$

This follows because

$$S^{n+1}(S^k(0)) = S^{n+k+1}(0) \neq S^k(0),$$

since $n + 1 + k \neq k$.

For a nonstandard number c, there is no way to prove that $S(c) \neq c$ from Robinson Arithmetic. When we assume that the Induction Principle holds in $\mathcal{M}$, this means that the conclusion of Example 6.3.2 holds in $\mathcal{M}$ as well and therefore $S(c) \neq c$ for nonstandard numbers as well.

Proposition 9.1.2. *It follows from Axioms (Q_1), (Q_2), (Q_3) and the Induction Principle for natural numbers that, for all n,*

$$(\forall x)S^{n+1}(x) \neq x.$$

Proof.   This is proved by fixing n and doing induction on x.

Base Step: $x = 0$. For any n, $S^{n+1}(0) = S(S^n(0)) \neq 0$ by Axiom (Q_1).

Induction Step: $x > 0$. Suppose that $S^{n+1}(x) \neq x$. Suppose by way of contradiction that $S^{n+1}(S(x)) = S(x)$. Then we have

$$S(S^{n+1}(x)) = S^{n+1}(S(x)) = S(x).$$

It follows from Axiom (Q_2) that $S^{n+1}(x)) = x$, which is a contradiction.   □

It is also possible to eliminate the finite orbits by adding the linear ordering. Recall from Section 3.4 the definition of a strict linear ordering $<$. The associated *weak* linear ordering is defined as $x \leq y \iff x < y \vee x = y$. The notion of successor is that $S(x)$ is the least element which is strictly greater than x. This may be given by the following.

$$(SL): (\forall x)(\forall y)(x < y \iff S(x) \leq y).$$

The structure $(\mathbb{N}, 0, S, \leq)$ satisfies the above axiom.

Example 9.1.3. Here is another structure which satisfies the linear ordering axioms as well as the first three axioms of Robinson Arithmetic. Let $A = \mathbb{N} \cup \{a_i : i \in \mathbb{Z}\}$. Define S on A by $S(n) = n+1$ for $n \in \mathbb{N}$ and $S(a_i) = a_{i+1}$ for $i \in \mathbb{Z}$. Define the ordering $<$ to be standard on $\mathbb{N}$, to have $a_i < a_j \iff i < j$ for all $i, j \in \mathbb{Z}$, and to have $n < a_i$ for all $i \in \mathbb{Z}$. It is also possible to have more orbits of type $\mathbb{Z}$, as we shall see later.

Proposition 9.1.4. *Any structure $\mathcal{M} = (M, S, 0, \leq)$ which is a linear ordering with successor and satisfies the first three axioms of Robinson Arithmetic has no finite orbits. It has one orbit of type ω and all other orbits of type $\mathbb{Z}$.*

Proof. It follows by induction on n that for any $x \in M$, $x < S^{n+1}(x)$.

Base Step: $n = 0$. For any x, $x < S(x)$ by (SL), since $S(x) \leq S(x)$.

Induction Step: $n > 0$. Suppose that $x < S^{n+1}(x)$. Then by the base step, $S^{n+1}(x) < S(S^{n+1}(x)) = S^{n+2}(x)$. Thus by transitivity, $x < S^{n+2}(x)$.

Thus it is not possible that $S^{n+1}(x) = x$, so there are no finite orbits. The orbit of 0 has type ω. Suppose that c is not in the orbit of 0. Then $c \neq S^n(0)$ for any n.

We can now show by induction that, for all n, there exists $a \in M$ such that $S^n(a) = c$. This implies that the orbit of c has type $\mathbb{Z}$.

Base Step: $n = 0$. $c = S^0(c)$.

Induction Step: $n > 0$. Suppose that $c = S^n(a)$ for some $a \in M$. Since c is not in the orbit of 0, $a \neq 0$. Thus by Axiom (Q_2), $a = S(a')$ for some $a' \in M$. Then $c = S^n(S(a')) = S^{n+1}(a')$. $\square$

Let us now consider the outcome of including addition as part of the language. Here are the two relevant axioms of Robinson Arithmetic.

$(Q_4):$ $(\forall x) x + 0 = x$ and

$(Q_5):$ $(\forall x)(\forall y)(x + S(y) = S(x + y))$

Example 9.1.5. We give a structure $\mathcal{M} = (M, S^{\mathcal{M}}, +^{\mathcal{M}}, <^{\mathcal{M}}, 0^{\mathcal{M}})$ which is a model of the theory of a linear ordering with successor and also satisfies the first five axioms of Robinson Arithmetic. Let $M = (\{0\} \times \mathbb{N}) \cup \{1, 2, \dots\} \times \mathbb{Z}$. Let $0^{\mathcal{M}} = (0, 0)$. Let $S(n, i) = (n, i+1)$. Let $(n_1, i_1) + (n_2, i_2) = (n_1 + n_2, i_1 + i_2)$. Let $(m, i) < (n, j)$ if and only if either $m < n$ or $m = n$ and $i < j$. It is easy to see that $\mathcal{M}$ has the desired properties. The orbit of $0^{\mathcal{M}}$ is $\{0\} \times \mathbb{N}$. For each n, the orbit of $(n, 0)$ is $\{n\} \times \mathbb{Z}$ is an orbit of type $\mathbb{Z}$.

Proposition 9.1.6. *Suppose that* $\mathcal{M} = (M, S, +, <, 0)$ *is a model of the theory of a linear ordering with successor and also satisfies the first five axioms of Robinson Arithmetic, and that $\mathcal{M}$ has an infinite element. Then $\mathcal{M}$ must have infinitely many orbits of type $\mathbb{Z}$.*

Proof. Let c be an infinite element of $\mathcal{M}$, so that $i < c$ for all $i \in \mathbb{N}$. Recall that

$$n \times c = \underbrace{c + c + \cdots + c}_{n \text{ times}},$$

and consider the orbits $\mathcal{O}(n \times c)$. For each $n \in \mathbb{N}$ and each $i > 0$, $n \times c < (n \times c) + i < (n \times c) + c = (n + 1) \times c$. It follows that the orbits $\mathcal{O}(n \times c)$ are all distinct. $\square$

For models of the Induction Principle, we can say more.

Proposition 9.1.7. *Suppose that* $\mathcal{M} = (M, S, +, <, 0)$ *is a model of the theory of a linear ordering with successor, satisfies the first five axioms of Robinson Arithmetic as well as the Induction Principle, and has an infinite element. Then the orbits of $\mathcal{M}$ are densely ordered. That is, for any elements a and b such that $a + n < b$ for all n, there exists c such that, for every n, both $a + n < c$ and $c + n < b$.*

Proof. The proof depends on the notions of odd and even numbers.

Lemma 9.1.8. *For any element $y \in \mathcal{M}$, there exists x such that either $y = x + x$ or $y = x + x + 1$.*

Proof. The proof is by the Induction Principle for $\mathcal{M}$.

Base Step: $y = 0$. Then $y = 0 + 0$ by Axiom (Q_4).

Induction Step: $y > 0$. Suppose that either $y = x + x$ or $y = x + x + 1$. There are two cases. If $y = x + x$, then $S(y) = y + 1 = x + x + 1$. If $y = x + x + 1$, then $S(y) = x + x + 2 = (x + 1) + (x + 1)$. Here we use the associative and commutative laws for addition, which can be proved by induction. □

Now suppose that a and b are elements such that $a + n < b$ for every n. By Lemma 9.1.8, there is an element c such that either $a + b = c + c$ or $a + b = c + c + 1$. If $a + b = c + c$, then for any n,

$$a + a + 2n < a + b = c + c,$$

so that $a + n < c$. Similarly if $a + b = c + c + 1$, then for any n,

$$a + a + 2n + 1 < a + b = c + c + 1,$$

and so again we have $a + n < c$. A similar argument shows that $c + n < b$ for all n. □

We observe that the argument above also shows that there is no least orbit of type $\mathbb{Z}$. It is left as an exercise to show that there is no greatest orbit.

Let U be an ultrafilter on $I = \mathbb{N}$ and let

$$\mathbb{N}^* = \bigotimes_{i \in I} \mathbb{N}/U.$$

Recall that the elements of $\mathbb{N}^*$ are equivalence classes $[x]$ of functions $x : I \to \mathbb{N}$ that agree on a set of values in U. It follows from Łoś's Theorem that $\mathbb{N}^* \equiv \mathbb{N}$. This means in particular that for any formula φ and any $x_1, \ldots, x_n$,

$$\mathbb{N}^* \models \varphi([x_1], \ldots, [x_n]) \iff \{i : \mathbb{N} \models \varphi(x_1(i), \ldots, x_n(i))\} \in U.$$

As for any ultrapower, the structure $\mathbb{N}$ may be embedded in $\mathbb{N}^*$ by mapping n to $[c_n]$ where $c_n(i) = n$ for all i.

Definition 9.1.9. Let us say that $x : I \to \mathbb{N}$ is *bounded* if for some n and all i, $x(i) \leq n$. Let us say that $[x]$ is a *finite* element of $\mathbb{N}^*$ if $[x] \leq [c_n]$ for some n; say that $[x]$ is *infinite* otherwise.

An example of an infinite nonstandard number is $[x]$, where $x(i) = i$ for all $i \in \mathbb{N}$. To check this, observe that for any n, and $i > n$, $c_n(i) = n < i = x(i)$. So $\{i : c_n(i) < x(i)\}$ is cofinite and hence in U. Thus $[c_n] < [x]$ for any n.

More generally, we have the following.

Proposition 9.1.10. *For any $x : I \to \mathbb{N}$, we have:*

(a) *If $\lim_{i \to \infty} x(i) = \infty$, then $[x]$ is an infinite element of $\mathbb{N}^*$.*
(b) *If x is bounded, then $[x]$ is finite.*

The proofs of these facts are left to the exercises.

Example 9.1.11. The implications of Proposition 9.1.10 cannot be reversed. Let

$$x(n) = \begin{cases} 0 & \text{if } n \text{ is even,} \\ n & \text{if } n \text{ is odd.} \end{cases}$$

Whether x is finite or infinite depends on the ultrafilter U. We know that either the set of even numbers belongs to U or the set of odd numbers belongs to U. In the first case, $[x] = [c_0]$ and is finite, whereas x is not bounded. In the second case, $[x]$ is infinite, whereas the limit $\lim_{i \to \infty} x(i)$ does not exist.

Next we consider the cardinality of the ultrapower $\mathbb{N}^*$. Certainly $\mathbb{N}^*$ is infinite, since it includes a copy of $\mathbb{N}$. But is it countable or uncountable?

Theorem 9.1.12. $\mathbb{N}^*$ *is uncountable.*

Proof. Suppose by way of contradiction that

$$\mathbb{N}^* = \{[x_n] : n \in \mathbb{N}\}.$$

We use a modified diagonal argument to create x such that $[x] \neq [x_n]$ for any n. Simply define $x(i) = 1 + \max\{x_n(i) : n \leq i\}$. Now fix n and observe that $x(i) > x_n(i)$ for all $i > n$. Since this occurs on a cofinite set, it follows that, for each n,

$$\{i : x_n(i) < x(i)\} \in U,$$

so that $[x_n] < [x]$ for every $n \in \mathbb{N}$. $\qquad\square$

Exercises for Section 9.1

Exercise 9.1.1. Verify that Example 9.1.5 is a linear order with successor and satisfies the first five axioms of Robinson Arithmetic.

Exercise 9.1.2. Give the details to verify that $c + n < b$ in the argument of Proposition 9.1.7.

Exercise 9.1.3. Show that under the conditions of Proposition 9.1.7 there is no greatest orbit.

Exercise 9.1.4. Show that if $<$ is a strict linear ordering as defined in Section 3.4, then the associated weak linear ordering $\leq$ is transitive, irreflexive, and antisymmetric in the sense that $a \leq b$ and $b \leq a$ imply that $a = b$.

Exercise 9.1.5. Show that, for any $x : I \to \mathbb{N}$, if $\lim_{i \to \infty} x(i) = \infty$, then $[x]$ is an infinite element of $\mathbb{N}^*$.

Exercise 9.1.6. Show that, for any $x : I \to \mathbb{N}$, if x is bounded, then $[x]$ is finite.

Exercise 9.1.7. Show that if $[x]$ is an infinite element of $\mathbb{N}^*$ and $\lim_i x(i)$ exists, then $\lim_{i \to \infty} x(i) = \infty$.

Exercise 9.1.8. Here is an alternative to Theorem 9.1.12. Define an embedding f of $2^{\mathbb{N}}$ into $\mathbb{N}^{\mathbb{N}}$ so that for $x \neq y$, if $x(n) \neq y(n)$ then $f(x)$ and $f(y)$ are different for all $m \geq n$. Show that $[f(x)]_U \neq [f(y)]_U$ for any ultrafilter U.

9.2.   Nonstandard Analysis

When Leibniz and Newton were developing the calculus, they did so without providing the rigorous mathematical foundation to which we have become accustomed. In particular, Leibniz worked in terms of infinitesimals, or numbers smaller in absolute value than any positive real number. The infinitesimals survive in the form of the notation dx used in integrals. An integral, $\int f(x)\,dx$, can be thought of as a continuous form of a sum of parts that consist of the area of the rectangle of height $f(x)$ and width dx. Nonstandard Analysis, or Infinitesimal Analysis, as it is sometimes called, is a theory that provides a foundation for Leibniz' idea of infinitesimals. Abraham Robinson [27] developed the modern approach to infinitesimals, showing that the use of infinitesimals is consistent with the usual notions of analysis and logic.

As with the structure $\mathbb{N}$ of natural numbers, it follows from the Compactness Theorem that there is a structure $\mathbb{R}^*$ which is elementarily equivalent to $\mathbb{R}$ but has an infinite element c. We will also want to add to the language names for each real number, which will make $\mathbb{R}$ an elementary submodel of $\mathbb{R}^*$. We also add names for real functions, so that we can discuss the notions of limits and derivatives. Thus each real function f will have an extension f^* to $\mathbb{R}^*$. We also add a relation R so that $\mathbb{R} = \{x \in \mathbb{R}^* : R(x)\}$.

As we did for $\mathbb{N}$ and $\mathbb{N}^*$, below we will construct $\mathbb{R}^*$ as an ultrapower. For this construction, the structure $\mathbb{R}^*$ may be viewed as an elementary extension of $\mathbb{R}$, by Corollary 4.4.7. Thus $\mathbb{R}^*$ is a field. Since the infinite element $c > 0$, it follows that c has a multiplicative inverse c^{-1}. Since $c > r$ for all standard reals r, it follows that $c^{-1} < r$ for all positive reals r. Elements of $\mathbb{R}^*$ are said to be *hyperreals*.

Definition 9.2.1.

(1) An element $x \in \mathbb{R}^*$ is *infinitesimal* if $|x| < r$ for all positive $r \in \mathbb{R}$; *finite* if $|x| < r$ for some $r \in \mathbb{R}$; *infinite* if $|x| > r$ for all $r \in \mathbb{R}$.
(2) Two elements $x, y \in \mathbb{R}^*$ are said to be *infinitely close*, in symbols $x \approx y$, if $x - y$ is infinitesimal.

Notice that the statement that x is infinitesimal is equivalent to the statement that $x \approx 0$.

Definition 9.2.2. Given a hyperreal number $x \in \mathbb{R}^*$, the *monad* of x is

$$\text{monad}(x) = \{y \in \mathbb{R}^* : x \approx y\}$$

and the *galaxy* of x is

$$\text{galaxy}(x) = \{y \in \mathbb{R}^* : x - y \text{ is finite}\}$$

The monad and galaxy picked out by 0 are of particular interest, since monad(0) is the set of infinitesimals and galaxy(0) is the set of finite hyperreals.

With the basic definitions and axioms in hand, we next turn to a discussion of the algebraic structure of the hyperreals.

Proposition 9.2.3.

(a) *The set* galaxy(0) *of finite elements of* $\mathbb{R}^*$ *is a subring of* $\mathbb{R}^*$ *(that is, it is closed under sums and the differences and products of finite elements are finite).*

(b) *The set* monad(0) *of infinitesimal elements is a subring of* $\mathbb{R}^*$ *and an ideal of* galaxy(0). *That is,*

   (i) *sums, differences and products of infinitesimals are infinitesimal;*
   (ii) *the product of an infinitesimal and a finite element is infinitesimal.*

Proof. Here is the proof of part (a). If $|x| < r$ and $|y| < s$, then $|x + y| < r + s$, $|x - y| < r + s$ and $|xy| < rs$.

Here is the proof of part (b). For part (i), suppose that $\epsilon, \delta \approx 0$. Then for each positive real number r, we have $|\epsilon| < r/2$ and $|\delta| < r/2$. Hence $|\epsilon + \delta| < r$ and $|\epsilon - \delta| < r$.

For part (ii), suppose ϵ is infinitesimal and b is finite. Let t be a real number so that $|b| < t$. Then for every positive real number r, we have $|\epsilon| < r/t$. So $|b\epsilon| < (r/t)t = r$. Thus $b\epsilon$ is infinitesimal. □

Corollary 9.2.4. *Any two galaxies are equal or disjoint.*

Proof. We use the *coset representation* of galaxies:

$$\text{galaxy}(x) = \{x + a \colon a \in \text{galaxy}(0)\}.$$

If $x + a = y + b$, then any element $x + c$ of galaxy(x) is in galaxy(y), since $x + c = x + a + (c - a) = y + b + (c - a)$. Similarly any element of galaxy(y) is in galaxy(x). □

Corollary 9.2.5. *Any two monads are equal or disjoint. Furthermore, the relation* $x \approx y$ *is an equivalence relation on* $\mathbb{R}^*$.

Proof. The proof is as above using the *coset representation* of monads:

$$\text{monad}(x) = \{x + \epsilon \colon \epsilon \in monad(0)\}.$$

Since $x \approx y$ if and only if monad$(x) = $ monad(y), the relation $\approx$ is an equivalence relation because $=$ is an equivalence relation. □

Proposition 9.2.6.

(1) $x \in \mathbb{R}$ *is infinite if and only if* x^{-1} *is infinitesimal.*
(2) $\mathrm{monad}(0)$ *is a maximal ideal in* $\mathrm{galaxy}(0)$. *(That is, there is no ideal* I *with* $\mathrm{monad}(0) \subsetneq I \subsetneq galaxy(0)$*).*

Proof. (1) An element x is infinite if and only if $|x| \geq r$ for all positive $r \in \mathbb{R}$. That condition holds if and only if $|x^{-1}| \leq r^{-1}$ for all positive $r \in \mathbb{R}$. Since as r ranges over all positive reals, r^{-1} ranges over all positive reals as well, the latter condition is equivalent to the statement that x^{-1} is infinitesimal.

(2) Suppose that $\mathrm{monad}(0) \subsetneq I \subseteq \mathrm{galaxy}(0)$. We show that $I = \mathrm{galaxy}(0)$. Let b be an element of I not in $\mathrm{monad}(0)$. Since $b = (b^{-1})^{-1}$ is not infinitesimal, from part (1) we know that b^{-1} is not infinite. Thus b^{-1} is in $\mathrm{galaxy}(0)$. So $1 = b \times b^{-1}$ is in I, since I is an ideal in $\mathrm{galaxy}(0)$. Thus any c in $\mathrm{galaxy}(0)$ is also in I, because we can represent it as $c = 1 \times c$. Therefore $I = \mathrm{galaxy}(0)$. $\qquad\square$

Proposition 9.2.7. *Every finite x in $\mathbb{R}^*$ is infinitely close to a unique $r \in \mathbb{R}$. That is, every finite monad contains a unique real number.*

Proof. Suppose $x \in \mathbb{R}^*$ is finite. First we show the existence of $r \in \mathbb{R}$.

Let $X = \{s \in \mathbb{R} : s < x\}$. Then X is nonempty and has an upper bound, since any real number t that demonstrates that x is finite is an upper bound for X. Thus X has a least upper bound r. For every positive real t, since r is an upper bound, we have

$$x \leq r + t \quad \text{or} \quad x - r \leq t.$$

Since r is the least upper bound of X, we also have

$$r - t \leq x \quad \text{or} \quad -(x - r) \leq t.$$

The two inequalities combine to give the inequality $|x - r| \leq t$. Since this inequality holds for every positive real t, the quantity $x - r$ is infinitesimal. That is, x is infinitely close to r.

We now show the uniqueness of r. Suppose that x is a hyperreal and that x is infinitely close to reals r and s. That is, $x \approx r$ and $x \approx s$. Then $r \approx s$, or $r - s \approx 0$. However, r and s are real, so their difference $r - s$ is also real. Thus $r - s = 0$ and so $r = s$.

□

Definition 9.2.8. Given a finite x in $\mathbb{R}^*$, the unique real r infinitely close to x is called the *standard part* of x, in symbols, $r = \text{st}(x)$. If x is infinite, $\text{st}(x)$ is undefined.

Corollary 9.2.9. *Let x and y be finite.*

(1) $x \approx y$ *if and only if* $\text{st}(x) = \text{st}(y)$.
(2) $x \approx \text{st}(x)$.
(3) *If* $r \in \mathbb{R}$, *then* $\text{st}(r) = r$.
(4) *If* $x \leq y$, *then* $\text{st}(x) \leq \text{st}(y)$.

Proof. The proof is left to the reader. □

Proposition 9.2.10.

(1) *There exist positive and negative infinitesimals in $\mathbb{R}^*$.*
(2) *$\mathbb{R}^*$ has positive and negative infinite elements.*

Proof. $\mathbb{R}^*$ has a positive infinite element c by its definition. Then $-c$ is a negative infinite element, $1/c$ is a positive infinitesimal and $-1/c$ is a negative infinitesimal. Thus there is a positive infinitesimal ϵ. From ϵ the other desired elements can be defined: $-\epsilon$ is a negative infinitesimal; $1/\epsilon$ is a positive infinite element; and $-1/\epsilon$ is a negative infinite element. □

We have seen that the product of an infinitesimal with a finite number results in an infinitesimal. In contrast, the product of an infinitesimal and an infinite number can be infinite, infinitesimal, or even finite. For example, suppose ϵ is an infinitesimal. Then ϵ^2 is also an infinitesimal, while $1/\epsilon$ and $1/\epsilon^2$ are both infinite.

Then:

$$\epsilon^2 \times \frac{1}{\epsilon} \quad \text{is infinitesimal;}$$

$$\epsilon \times \frac{1}{\epsilon} \quad \text{is finite, not infinitesimal;}$$

$$\epsilon \times \frac{1}{\epsilon^2} \quad \text{is infinite.}$$

Proposition 9.2.11. *The standard part function is a homomorphism of the ring* galaxy(0) *onto the field of real numbers. That is, for finite real numbers x and y,*

(1) $\text{st}(x + y) = \text{st}(x) + \text{st}(y)$;
(2) $\text{st}(x - y) = \text{st}(x) - \text{st}(y)$;
(3) $\text{st}(x \times y) = \text{st}(x) \times \text{st}(y)$.

Proof. Parts (1) and (2) are left to the reader. Suppose x and y are finite, and let r be the standard part of x and s the standard part of y. Then $x = r + \epsilon$ and $y = s + \delta$ for some infinitesimals ϵ and δ. The standard part of $x \times y$ is $\text{st}((r + \epsilon)(s + \delta)) = \text{st}(rs + r\delta + s\epsilon + \epsilon\delta)$. Since $r\delta$, $s\epsilon$ and $\epsilon\delta$ are all infinitesimal, it follows that the standard part of $x \times y$ is rs. That is, part (3) follows. □

Corollary 9.2.12. *Let x and y be finite.*

(1) *If* $\text{st}(y) \neq 0$, *then* $\text{st}\left(\frac{x}{y}\right) = \frac{\text{st}(x)}{\text{st}(y)}$.
(2) *If* $y = \sqrt[n]{x}$, *then* $\text{st}(y) = \sqrt[n]{\text{st}(x)}$.

We now want to discuss limits and derivatives of function on reals and hyperreals. It is important here that the real functions $f : \mathbb{R} \to \mathbb{R}$ to be discussed will have a name in the language, and so will have a natural extension $f^* : \mathbb{R}^* \to \mathbb{R}^*$. There is an excellent presentation of elementary calculus using infinitesimals by Keisler [21].

Recall some basic definitions from calculus.

Definition 9.2.13.

(1) $\lim_{x \to a} f(x) = v$ if and only if for every $\epsilon > 0$, there exists $\delta > 0$ such that, for all $x \neq a$, $|x - a| < \delta$ implies $|f(x) - f(a)| < \epsilon$.
(2) A function $f : \mathbb{R} \to \mathbb{R}$ is *continuous* at the point $x = a$ if $\lim_{x \to a} f(x) = f(a)$.
(3) A function $f : \mathbb{R} \to \mathbb{R}$ is *differentiable* at the point $x = a$ and the derivative $f'(a) = d$ if $\lim_{x \to a} \frac{f(x)-b}{x-a} = d$.

What happens when we allow infinitesimal ϵ and δ? It is important to recall here that $\mathbb{R}$ may be viewed as an elementary submodel of $\mathbb{R}^*$. That is, for any formula φ and any elements $a_1, \ldots, a_m \in \mathcal{A}$,

$$\mathcal{A} \models \varphi(a_1, \ldots, a_m) \iff \mathcal{A}^* \models \varphi([c_{a_1}], \ldots, [c_{a_n}]),$$

where we identify the constant function c_a with a in $\mathbb{R}^*$.

Proposition 9.2.14. *Let $f : \mathbb{R} \to \mathbb{R}$ have extension $f^* : \mathbb{R}^* \to \mathbb{R}^*$ and let $a, v \in \mathbb{R}$. Then $\lim_{x \to a} f(x) = v$ in $\mathbb{R}$ if and only if, for all $x \in \mathbb{R}^*$ with $x \neq a$, if $\mathrm{st}(x) = a$, then $\mathrm{st}(f^*(x)) = v$.*

Proof. Suppose first that in $\mathbb{R}$, $\lim_{x \to a} f(x) = v$ and let $x - a \neq 0$ be infinitesimal. Now let $\epsilon > 0$ be in $\mathbb{R}$. Then there is some $\delta > 0$ in $\mathbb{R}$ such that $|x - a| < \delta$ implies $|f(x) - v| < \epsilon$. It follows that in $\mathbb{R}^*$, $|x - a| < \delta$ implies $|f^*(x) - v| < \epsilon$. Since $x - a$ is infinitesimal, $|x - a| < \delta$ and therefore $|f^*(x) - v| < \epsilon$. Since this is true for all positive reals ϵ, it follows that $f^*(x) - v$ is infinitesimal and hence $\mathrm{st}(f^*(x)) = v$.

Next suppose that, for every real $x \neq a$, if $\mathrm{st}(x) = a$ then $\mathrm{st}(f^*(x)) = v$. Let the real $\epsilon > 0$ be given and let δ be any infinitesimal. Then $|x - a| < \delta$ implies that $\mathrm{st}(x) = a$ and therefore $\mathrm{st}(f^*(x)) = v$, so that $|f^*(x) - v| < \epsilon$ since $f^*(x) - v$

is infinitesimal. Thus $\mathbb{R}^*$ satisfies the formula

$$\varphi(a, v, \epsilon) : (\exists \delta > 0)(\forall x \neq a)[|x - a| < \delta \rightarrow |f^*(x) - v| < \epsilon].$$

It follows that $\mathbb{R} \models \varphi(a, v, \epsilon)$ as well, so that $\lim_{x \to a} f(x) = v$.

$\square$

We observe that for a hyperreal x and real a, $\text{st}(x) = a$ if and only if $x \approx a$. Thus $\lim_{x \to a} f(x) = b$ if and only if, for all $x \neq a$, if $x \approx a$ then $f^*(x) \approx b$.

Corollary 9.2.15. *A function $f : \mathbb{R} \to \mathbb{R}$ is continuous at the point $(a, f(a))$ if for all $x \approx a$, $f^*(x) \approx f(a)$.*

Next we examine derivatives of real functions. Recall that the derivative $f'(a) = \lim_{\delta \to 0} \frac{f(a+\delta)-f(a)}{\delta}$, if this limit exists. If $f'(a)$ exists, then we say that f is *differentiable* at $x = a$. Thus we have the following corollary.

Corollary 9.2.16. *For any real function f and any reals a and d, $f'(a) = b$ if and only if, for all nonzero infinitesimals δ, $\text{st}\left(\frac{f(a+\delta)-f(a)}{\delta}\right) = b$.*

Example 9.2.17. Let $f(x) = x^2$. Then for any a and δ,

$$\frac{f(a+\delta) - f(a)}{\delta} = \frac{(a+\delta)^2 - a^2}{\delta} = 2a + \delta.$$

For any infinitesimal nonzero δ, it follows that $\text{st}\left(\frac{f(a+\delta)-f(a)}{\delta}\right) = \text{st}(2a + \delta) = 2a$.

Basic results about derivatives and continuity are a bit easier to prove using infinitesimals.

Proposition 9.2.18. *If a function f is differentiable at the point $x = a$, then it is continuous at $x = a$.*

Proof. Suppose that $f'(a) = d$, so for any infinitesimal nonzero δ, it follows that $\mathrm{st}\left(\frac{f(a+\delta)-f(a)}{\delta}\right) = d$. Then $\mathrm{st}(f(a+\delta) - f(a)) = \mathrm{st}(d\delta) = 0$, so that $f(a+\delta) \approx f(a)$. □

Proposition 9.2.19. *Suppose that the functions f and g are both differentiable at $x = a$, and let c be any real. Then the functions cf, $f+g$, and $f \times g$ are differentiable at $x = a$ and the function f/g is differentiable at $x = a$ if $g(a) \neq 0$. Moreover, we have the following formulas:*

(a) $(cf)'(a) = cf'(a)$;
(b) $(f+g)'(a) = f'(a) + g'(a)$;
(c) $(f \times g)'(a) = f'(a)g(a) + f(a)g'(a)$;
(d) $(f/g)'(a) = \frac{f'(a)g(a)-f(a)g'(a)}{g(a)^2}$.

Proof. For part (a), Suppose that $f'(a) = d$. Then for any nonzero infinitesimal δ,

$$\mathrm{st}\left(\frac{f(a+\delta) - f(a)}{\delta}\right) = d,$$

and it follows that

$$\mathrm{st}\left(\frac{cf(a+\delta) - cf(a)}{\delta}\right) = cd,$$

For part (d), let $\delta \neq 0$ be infinitesimal. We have $\frac{f(a+\delta)-f(a)}{\delta} \approx f'(a)$ and $\frac{g(a+\delta)-g(a)}{\delta} \approx g'(a)$. Then

$$\left(\frac{f(a+\delta)}{g(a+\delta)} - \frac{f(a)}{g(a)}\right)\Big/\delta = \frac{f(a+\delta) - f(a)}{\delta g(a+\delta)g(a)} \cdot g(a)$$

$$- f(a) \cdot \frac{g(a+\delta) - g(a)}{\delta g(a+\delta)g(a)}$$

$$\approx \frac{f'(a)g(a) - f(a)g'(a)}{g(a)^2}.$$

Parts (b) and (c) are left to the reader. □

Finally we consider the Chain Rule.

Theorem 9.2.20. *Suppose that $f(a) = b$, that f is differentiable at a and g is differentiable at b and let h be the composition $g \circ f$. Then h is differentiable at a and $h'(a) = g'(b)f'(a)$.*

Proof. Let $\delta \neq 0$ be infinitesimal. Since f is continuous at a by the assumption that f is differentiable at a and Proposition 9.2.18, it follows that $f(a + \delta) \approx f(a)$. Thus $f(a + \delta) = f(a) + \epsilon = b + \epsilon$ for some infinitesimal ϵ. Note that $\epsilon = f(a + \delta) - f(a)$, so that $\frac{f(a+\delta)-f(a)}{\epsilon} = 1$. Now we have $\frac{g(b+\epsilon)-g(b)}{\epsilon} \approx g'(b)$ and $\frac{f(a+\delta)-f(a)}{\delta} \approx f'(a)$. Then

$$\frac{h(a+\delta) - h(a)}{\delta} = \frac{g(f(a+\delta)) - g(f(a))}{\delta} = \frac{g(b+\epsilon) - g(b)}{\delta}$$

$$= \frac{g(b+\epsilon) - g(b)}{\delta} \frac{f(a+\delta) - f(a)}{\epsilon}$$

$$= \frac{g(b+\epsilon) - g(b)}{\epsilon} \frac{f(a+\delta) - f(a)}{\delta} \approx g'(b)f'(a).$$

$\square$

Let U be an ultrafilter on $I = \mathbb{N}$ and let

$$\mathbb{R}^* = \bigotimes_{i \in I} \mathbb{R}/U.$$

Recall that the elements of $\mathbb{R}^*$ are equivalence classes $[x]$ of functions $x : I \to \mathbb{R}$. It follows from of Łoś's theorem that $\mathbb{N}^* \approx \mathbb{N}$, since, for any formula φ and any $x_1, \ldots, x_n$:

$$\mathbb{R}^* \models \varphi([x_1], \ldots, [x_n]) \iff \{i : \mathbb{R} \models \varphi(x_1(i), \ldots, x_n(i))\} \in U.$$

As for any ultrapower, the structure $\mathbb{R}$ may be embedded in $\mathbb{R}^*$ by mapping r to $[c_r]$ where $c_r(i) = r$ for all i. We will generally denote c_r by r. Then Corollary 4.4.7 tells us that for any formula φ and any reals $a_1, \ldots, a_n$,

$$\mathbb{R} \models \varphi(a_1, \ldots, a_m) \iff \mathbb{R}^* \models \varphi([a_1], \ldots, [a_n]).$$

Every nonstandard natural number is an infinite hyperreal. A hyperreal x is infinite if and only if it is $x > r$ for all standard reals r, which is equivalent to $x > n$ for all standard natural numbers n. Then a hyperreal is finite if and only if $x < r$ for some standard r, or equivalently, $x < n$ for all standard natural numbers n. Recall that a finite element of $\mathbb{N}^*$ is simply a standard natural number. In $\mathbb{R}^*$, the hyperreal x is finite if and only if $x \approx r$ for some standard real r.

As with functions $x : I \to \mathbb{N}$, we say that $x : I \to \mathbb{R}$ is *bounded* if, for some n and all i, $x(i) \leq n$; this is equivalent to saying that for some standard real r and all i, $x(i) \leq r$. Then Proposition 9.1.10 can be extended as follows.

Proposition 9.2.21. *For any $x : I \to \mathbb{R}$,*

(a) *if $\lim_{i \to \infty} x(i) = \infty$, then $[x]$ is an infinite element of $\mathbb{R}^*$;*
(b) *if x is bounded, then $[x]$ is finite.*

Proof. Part (a) is left to the exercises. For part (b), suppose that x is bounded. Fix an ultrafilter U and let S be the set of reals r such that $\{i : x(i) \leq r\} \in U$. S is not empty since if $x(i) \leq r$ for all i, then $\{i : x(i) \leq r\} = I \in U$. Now let s be the supremum of S. It can be checked that $[x] \approx s$. $\qquad\square$

Recall from Theorem 9.1.12 that $\mathbb{N}^*$ is always uncountable, so it is larger than $\mathbb{N}$. It is natural to ask whether $\mathbb{R}^*$ is larger than $\mathbb{R}$. Recall from basic set theory that the cardinality card($\mathbb{R}$) of the reals is the same as the cardinality card($2^{\mathbb{N}}$) of the set of functions from $\mathbb{N}$ to $\{0, 1\}$. (See the companion volume on set theory [4] for this and other results on cardinality). It can be shown that $\mathbb{R}^*$ has the same cardinality as $\mathbb{R}$.

Exercises for Section 9.2

Exercise 9.2.1. Let $x, y \in \mathbb{R}^*$ be finite. Show the following:

(1) $x \approx y$ if and only if st$(x) = $ st(y).
(2) $x \approx $ st(x).

(3) If $r \in \mathbb{R}$, then $\mathrm{st}(r) = r$.

(4) If $x \leq y$, then $\mathrm{st}(x) \leq \mathrm{st}(y)$.

Exercise 9.2.2. Show that, for finite hyperreal numbers x and y,

(1) $\mathrm{st}(x + y) = \mathrm{st}(x) + \mathrm{st}(y)$;

(2) $\mathrm{st}(x - y) = \mathrm{st}(x) - \mathrm{st}(y)$.

Exercise 9.2.3. Complete the proof of Proposition 9.2.21 by showing that $[x] \approx s$.

Exercise 9.2.4. Use infinitesimals to show that for $f(x) = x^3$, $f'(a) = 3a^2$ for any real a.

Exercise 9.2.5. Suppose that the functions f and g are both differentiable at $x = a$. Show that the functions $f + g$ and $f \times g$ are differentiable at $x = a$ and that

(a) $(f + g)'(a) = f'(a) + g'(a)$;

(b) $(f \times g)'(a) = f'(a)g(a) + f(a)g'(a)$.

Chapter 10

Foundations of Geometry

Plane geometry is an area of mathematics that has been studied since ancient times. The roots of the word *geometry* are the Greek words *ge*, meaning "earth", and *metria*, meaning "measuring". This name reflects the computational approach to geometric problems, that had been used before the time of Euclid, (ca. 300 BC), who introduced the axiomatic approach in his book, *Elements*. Euclid did not limit himself to plane geometry in the *Elements*, but also included chapters on algebra, ratio, proportion and number theory. His book set a new standard for the way mathematics was done.

10.1. Axioms of Plane Geometry

Definition 10.1.1. The language of *Plane Geometry*, PG, has two one-place predicates, Pt and Ln, to distinguish the two kinds of objects in plane geometry, and a binary *incidence* relation, In, to indicate that a point is *on*, or incident with a line. We will sometimes write $P \in Pt$ for $Pt(P)$ and $\ell \in Ln$ for $Ln(\ell)$. As shorthand, we will also write $P, Q \ In \ \ell$ to indicate that both P and Q are incident with the line ℓ.

Thus a *model* $\mathcal{G}$ of plane geometry will consist of a set of points, a set of lines, and a binary incidence relation; we will say that such a model is a *geometry*.

We will begin with the following five axioms, roughly based on Hilbert's approach in his classic book, *Foundations of Geometry* [16].

Definition 10.1.2. A geometry $\mathcal{G}$ is a model of PG if it satisfies the following axioms:

(A_0) (Everything is either a point or line, but not both; only points are on lines).
$(\forall x)((x \in Pt \vee x \in Ln) \wedge \neg(x \in Pt \wedge x \in Ln)) \wedge$
$(\forall x, y)(x\ In\ y \rightarrow (x \in Pt \wedge y \in Ln)).$

(A_1) (Any two points belong to a line).
$(\forall P, Q \in Pt)(\exists \ell \in Ln)(P\ In\ \ell \wedge Q\ In\ \ell).$

(A_2) (Every line has at least two points).
$(\forall \ell \in Ln)(\exists P, Q \in Pt)(P\ In\ \ell \wedge Q\ In\ \ell \wedge P \neq Q).$

(A_3) (Two lines intersect in at most one point).
$(\forall \ell, g \in Ln)(\forall P, Q \in Pt)((\ell \neq g \wedge P, Q\ In\ \ell \wedge P, Q\ In\ g) \rightarrow P = Q).$

(A_4) (There are four points no three on the same line).
$(\exists P_0, P_1, P_2, P_3 \in Pt)(P_0 \neq P_1 \neq P_2 \neq P_3 \wedge P_2 \neq P_0 \neq P_3 \neq P_1$
$\wedge (\forall \ell \in Ln)(\neg(P_0\ In\ \ell \wedge P_1\ In\ \ell \wedge P_2\ In\ \ell) \wedge$
$\neg(P_0\ In\ \ell \wedge P_1\ In\ \ell \wedge P_3\ In\ \ell) \wedge$
$\neg(P_0\ In\ \ell \wedge P_2\ In\ \ell \wedge P_3\ In\ \ell) \wedge$
$\neg(P_1\ In\ \ell \wedge P_2\ In\ \ell \wedge P_3\ In\ \ell)).$

The axiom A_0 simply says that our objects have the types we intend, and is of a different character than the other axioms. In addition to these axioms, Euclid had one that asserted the existence of circles of arbitrary center and arbitrary radius, and one that asserted that all right angles are equal. He also had another axiom for points and lines, called the parallel postulate, which he attempted to show was a consequence of the other axioms.

Definition 10.1.3. Two lines are *parallel* if there is no point incident with both of them.

Definition 10.1.4. For $n \geq 0$, the *n-parallel postulate*, P_n, is the following statement:

(P_n) For any line ℓ and any point Q not on the line ℓ, there are exactly n lines parallel to ℓ through the point Q.

Consider also the axiom schema $P_{\geq n}$, as follows.

$(P_{\geq n})$ For any line ℓ and any point Q not on the line ℓ, there are at least n lines parallel to ℓ through the point Q.

Lastly, we have:

(P_∞) $(\forall n)P_{\geq n}$.

A geometry satisfies P_∞ if and only if, for any line ℓ and any point Q not on the line ℓ, there are infinitely many lines parallel to ℓ through the point Q.

P_1 is the famous parallel postulate. For nearly two thousand years, people tried to prove what Euclid had conjectured. Namely, they tried to prove that P_1 was a consequence of the other axioms. In the 1800s, models of the other axioms were produced which were not models of P_1.

10.2. Non-Euclidean Models

Nikolai Lobachevski (1793–1856), a Russian mathematician, and Janos Bolyai (1802–1860), a Hungarian mathematician, both produced models of the other axioms together with the parallel postulate P_∞. This geometry is known as *Lobachevskian geometry*.

Example 10.2.1 (Fig. 10.1). Fix a circle C in a Euclidean plane. The points of the geometry are the interior points of C.

Fig. 10.1.   A model of Lobachevskian geometry.

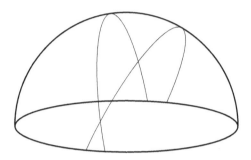

Fig. 10.2.   A model of Riemannian geometry.

The lines of the geometry are the intersection of lines of the Euclidean plane with the interior of the circle. Given any line ℓ of the geometry and any point Q of the geometry which is not on ℓ, every Euclidean line through Q which intersects ℓ on or outside of C gives rise to a line of the geometry which is parallel to ℓ.

Example 10.2.2 (Fig. 10.2). Fix a sphere S in Euclidean 3-space. The points of the geometry may be thought of as either the points of the upper half of the sphere, or as equivalence classes consisting of the pairs of points on opposite ends of diameters of the sphere (antipodal points). If one chooses to look at the points as coming from the upper half of the sphere, one must take care to get exactly one from each of the equivalence classes. The lines of the geometry are the intersection of the great circles with the points. Since any two great circles meet in two antipodal points, every pair of lines intersects. Thus this model satisfies P_0. This is a model of *Riemannian geometry*.

10.3. Finite Geometries

Next we turn to finite geometries, ones with only finitely many points and lines.

Definition 10.3.1. A geometry $\mathcal{G}$ is a projective plane of order q if it satisfies the five axioms of PG as well as the following:

$(A_5(q))$: Every line contains exactly $q+1$ points.
$(A_6(q))$: Every point lies on exactly $q+1$ lines.

We will write PG(q) as the theory PG along with the axioms $(A_5(q))$ and $(A_6(q))$.

It is not hard to see that there can be no projective plane of order 1. This is left as an exercise. The first geometry we look at is the finite projective plane of order 2, PG(2), also known as the Fano plane (Fig. 10.3).

Theorem 10.3.2. *The theory* PG(2) *has a unique model (up to isomorphism), which is a finite geometry of seven points and seven lines.*

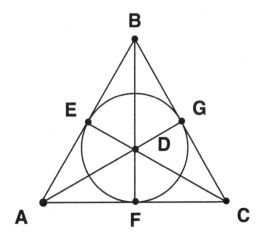

Fig. 10.3. The Fano plane.

Proof. See Exercise 10.3.4 to prove from the axioms and Exercise 10.3.3 that the following diagram gives a model of PG(2) and that any model must have the designated number of points and lines.   □

We can also construct models of projective geometry using vector spaces. The vector space underlying the construction is the vector space of dimension three over the field of two elements, $\mathbb{Z}_2 = \{0, 1\}$. The points of the geometry are one-dimensional subspaces. Since a one-dimensional subspace of $\mathbb{Z}_2$ has exactly two triples in it, one of which is the triple $(0, 0, 0)$, we identify the points with the triples of 0's and 1's that are not all zero. So the points are $P_{(1,0,0)}$, $P_{(0,1,0)}$, $P_{(0,0,1)}$, $P_{(1,1,0)}$, $P_{(1,0,1)}$, $P_{(0,1,1)}$, and $P_{(1,1,1)}$. In the incidence table below, point $P_{(i,j,k)}$ is shortened to just (i, j, k). The lines of the geometry are the two-dimensional subspaces. The incidence relation is defined so that a point P is on a line ℓ if the one-dimensional subspace (point) is a subspace of the two-dimensional subspace (line). Since a two-dimensional subspace may be viewed as the orthogonal complement of a one-dimensional subspace, each two-dimensional subspace ℓ is identified in the incidence table below with the nonzero triple $[a, b, c]$ such that $(i, j, k) \in \ell \iff (a, b, c) \cdot (i, j, k) \equiv 0 \pmod 2$. For example, the line $[1, 0, 0]$ is the two-dimensional space consisting of $\{(0, 0, 0), (0, 1, 0), (0, 0, 1), (0, 1, 1)\}$.

In	$[1, 0, 0]$	$[0, 1, 0]$	$[0, 0, 1]$	$[1, 1, 0]$	$[1, 0, 1]$	$[0, 1, 1]$	$[1, 1, 1]$
$(1, 0, 0)$	0	1	1	0	0	1	0
$(0, 1, 0)$	1	0	1	0	1	0	0
$(0, 0, 1)$	1	1	0	1	0	0	0
$(1, 1, 0)$	0	0	1	1	0	0	1
$(1, 0, 1)$	0	1	0	0	1	0	1
$(0, 1, 1)$	1	0	0	0	0	1	1
$(1, 1, 1)$	0	0	0	1	1	1	0

This table illustrates that the model has exactly seven points and seven lines, and that each point lies on exactly three lines and each line contains exactly three points. It can be seen that this model is isomorphic to the Fano plane presented above. See the exercises.

The vector space construction works over other finite fields as well. Consider for example the vector space V of dimension 3 over the field of three elements, $\mathbb{Z}_3 = \{0, 1, 2\}$. To construct a model $\mathcal{G}$ of PG(3), we can let the points of $\mathcal{G}$ be the one-dimensional subspaces of V and the lines can be the two-dimensional subspaces. The vector space V has 27 elements, so there are 13 one-dimensional subspaces each containing two nonzero vectors. As above, the lines are the orthogonal complements of the subspaces that form the lines, so there are also 13 of them.

This construction works for each finite field. In each case the order of the projective geometry is the size of the field.

The next few lemmas list a few facts about projective planes. Proofs are left to the exercises.

Lemma 10.3.3. *In any model of* PG(q), *any two lines intersect in a point.*

Lemma 10.3.4. *In any model of* PG(q), *there are exactly $q^2 + q + 1$ points.*

Lemma 10.3.5. *In any model of* PG(q), *there are exactly $q^2 + q + 1$ lines.*

For models built using the vector space construction over a field of q elements, it is easy to compute the number of points and lines as $\frac{q^3-1}{q-1} = q^2 + q + 1$. However, it has been shown that there are non-isomorphic projective planes of the same order. In particular, there are exactly four non-isomorphic planes of order nine.

Exercises for Chapter 10

Exercise 10.3.1. Translate Axioms $A_5(q)$ and $A_6(q)$ into the formal language of plane geometry.

Exercise 10.3.2. Define an isomorphism between the two models presented above of PG(2).

Exercise 10.3.3. Prove from the axioms that any two lines in PG(q) must intersect in a point. (*Hint*: Show that if g and h do not intersect and P is incident with g, then P is on at least one more line than h has points).

Exercise 10.3.4. Construct a model for PG(2) starting with four noncollinear points A, B, C and D and denoting the additional point on the line $\overline{AB}$ by E, the additional point on the line $\overline{AC}$ by F, and the additional point on the line $\overline{BC}$ by G. Use the axioms and Exercise 10.3.3 to justify the construction.

Exercise 10.3.5. Prove from the axioms that in any model of PG(q) there are exactly $q^2 + q + 1$ points.

Exercise 10.3.6. Prove from the axioms that in any model of PG(q) there are exactly $q^2 + q + 1$ lines.

Exercise 10.3.7. Show that there can be no projective plane of order 1.

Exercise 10.3.8. List the 13 one-dimensional subspaces of $\mathbb{Z}_3^3$ by giving one generator of each. (*Hint*: Proceeding lexicographically, four of them begin with "0" and the other nine begin with "1"). These are the points of PG(3). Identify the 13 two-dimensional subspaces of $\mathbb{Z}_3^3$ as orthogonal complements of these one-dimensional spaces. These are the lines of PG(3). For each line, list the four points on the line.

Exercise 10.3.9. Show that the axioms $(A_1), (A_2), (A_3), (A_4)$ for Plane Geometry are independent by constructing models which satisfy exactly 3 of the axioms, as well as axiom (A_0). (There are four possible cases here).

Bibliography

[1] J. Bell and A. Slomson. *Models and Ultraproducts*. North-Holland, 1969.

[2] A. Bezboruah and J. C. Shepherdson. Gödel's second incompleteness theorem for Q. *J. Symbolic Logic*, 41:503–512, 1976.

[3] G. Boole. *The Mathematical Analysis of Logic*. MacMillan, 1847.

[4] D. Cenzer, J. Larson, C. Porter and J. Zapletal. *Set Theory and the Foundations of Mathematics: An Introduction to Mathematical Logic — Volume 1: Set Theory*. World Scientific, 2020.

[5] G. J. Chaitin. Information-theoretic limitations of formal systems. *J. ACM (JACM)*, 21(3):403–424, 1974.

[6] C. C. Chang and H. J. Keisler. *Model Theory*, Third Edition, Studies in Logic and the Foundations of Mathematics, Vol. 73. Elsevier, 1990.

[7] B. Cooper. *Computability Theory*. CRC Press, 2003.

[8] M. Davis (ed.) *The Undecidable, Basic Papers on Undecidable Propositions, Unsolvable Problems And Computable Functions*. Raven Press, 1965.

[9] A. De Morgan. *Formal Logic*. Taylor, Walton and Malbery, 1847.

[10] R. Downey and D. Hirschfeldt. *Algorithmic Randomness and Complexity*. Springer, 2016.

[11] H. Enderton. *A Mathematical Introduction to Logic*. Academic Press, 1970.

[12] G. Frege. *A Formal Language for Pure Thought Modeled on that of Arithmetic*. Halle, 1879.

[13] K. Gödel. *Über die Vollständigkeit des Logikkalküls*. Ph.D. thesis, University of Vienna, 1929.

[14] L. Henkin. The completeness of the first-order functional calculus. *J. Symbolic Logic*, 14:159–166, 1949.

[15] D. Hilbert. Mathematical problems. *Bull. Amer. Math. Soc.*, 8:437–479, 1902.

[16] D. Hilbert. *Foundations of Geometry*. Open Court, 1971. Translated from German by Leo Unger.

[17]   D. Hilbert and W. Ackermann. *Grundzüge der Theoretischen Logik.* Springer, 1928.

[18]   P. G. Hinman. *Fundamentals of Mathematical Logic.* CRC Press, 2018.

[19]   W. Hodges. *A Shorter Model Theory.* Cambridge University Press, 1997.

[20]   J. Hopcroft, R. Motwani and J. Ullman. *Introduction to Automata Theory, Languages and Computation.* Third Edition. Pearson, 2007.

[21]   H. Jerome Keisler. *Elementary Calculus: An Infinitesimal Approach.* Prindle, Weber and Schmidt, 1976.

[22]   A. N. Kolmogorov. Three approaches to the definition of the concept "quantity of information". *Problemy Peredachi Informatsii,* 1(1):3–11, 1965.

[23]   M. Li and P. Vitanyi. *An Introduction to Kolmogorov Complexity and Its Applications.* Springer, 2008.

[24]   L. Löwenheim. Über möglichkeiten im relativkalkül. *Math. Ann.,* 76: 447–470, 1915.

[25]   D. Marker. *Model Theory.* Springer, 2002.

[26]   A. Nies. *Computability and Randomness.* Oxford University Press, 2009.

[27]   A. Robinson. *Non-standard Analysis.* Princeton University Press, 1974.

[28]   R. Sikorski. *Boolean Algebras.* Springer-Verlag, 1969.

[29]   M. Sipser. *Introduction to the Theory of Computation.* Third Edition, Cengage, 2012.

[30]   R. Soare. *Turing Computability.* Springer, 2016.

[31]   R. R. Stoll. *Set Theory and Logic.* W.H. Freeman, 1979.

[32]   P. Suppes. *Introduction to Logic.* Van Nostrand, 1957.

[33]   A. Tarski. *A Decision Method for Elementary Algebra and Geometry.* University of California Press, Berkeley, CA, 1951.

Index

A

absorption, 119
accepting, 148
alphabet, 5
arithmetic, 86
    Peano, 179, 181
    Robinson, 178
assignment, 58
associative, 116
atom, 121
axiomatizable, 43, 99
    finite, 99
    finitely, 43

B

Boole, George, 5, 115
Boolean algebra, 94, 115
    atomic, 122
    atomless, 122
    dense, 122
    interval, 117
    Lindenbaum, 117
bounded quantification, 163

C

cases, 161
categorical, 106
chain, 102
    union of a chain, 102
Chaitin, Gregory, 200
characteristic function, 138
Church's Theorem, 186
Church, Alonzo, 137, 186
Church–Turing thesis, 139
commutative, 90, 116
compactness, 32
    topological, 46
Compactness Theorem, 38, 86
complement, 116
complete, 37
Completeness Theorem, 37, 77–78
computably enumerable, 170–171
concatenation, 139
configuration, 148
connective, 5, 8, 13, 21–22